Environment and Technology in the Former USSR

NEW HORIZONS IN ENVIRONMENTAL ECONOMICS

General Editor: Wallace E. Oates, *Professor of Economics, University of Maryland*

This important series is designed to make a significant contribution to the development of the principles and practices of environmental economics. It includes both theoretical and empirical work. International in scope, it addresses issues of current and future concern in both East and West and in developed and developing countries.

The main purpose of the series is to create a forum for the publication of high quality work and to show how economic analysis can make a contribution to understanding and resolving the environmental problems confronting the world in the late twentieth century.

Recent titles in the series include:

Environment and Technology in the Former USSR

The Case of Acid Rain and Power Generation

Malcolm R Hill
Professor of Russian and East European Industrial Studies, Loughborough University, UK

Edward Elgar
Cheltenham, UK • Lyme, US

Published by
Edward Elgar Publishing Limited
8 Lansdown Place
Cheltenham
Glos GL50 2HU
UK

Edward Elgar Publishing, Inc.
1 Pinnacle Hill Road
Lyme
NH 03768
US

A catalogue record for this book
is available from the British Library

Library of Congress Cataloguing-in-Publication Data
Hill, Malcolm R.
    Environment and technology in the former USSR : the case of acid rain and power generation / by Malcolm R. Hill.
        (New horizons in environmental economics)
    Includes index.
        1. Air—Pollution—Former Soviet republics. 2. Fossil fuel power plants—Environmental aspects—Former Soviet republics. 3. Acid rain—Former Soviet republics. 4. Electric power production—Technology transfer—Former Soviet republics. 5. Technical assistance—Former Soviet republics. I. Title. II. Series.
    HC340.12.Z9A445 1997
    363.739'2'0947—dc21   97–16473
                                                                              CIP

ISBN 1 85898 583 8

Printed and bound in Great Britain by
Biddles Limited, Guildford and King's Lynn

Contents

List of Figures

List of Tables

Acknowledgements

The author wishes to gratefully acknowledge the financial support of the Global Environmental Change Research Programme of the Economic and Social Research Council, through the award of a Research Fellowship during the 1994-95 academic year. The award of this Fellowship provided the necessary time and travel funds to enable the author to research and analyse much of the material contained in this book.

The project was also enhanced through the award of a British Council Training and Academic Link to the author for the development of joint training programmes in power engineering, with Professor Yu. M. Porhovnik of the Management Development Centre at Saint Petersburg State Engineering Economics Academy. This link assisted in the setting up of discussions between the author and senior executives from the power engineering industry in the Saint Petersburg region, particularly from the *Leningradskii Metallicheskii Zavod* (*LMZ*) turbine factory, the *Leningradskii Zavod Turbinnykh Lopatok* (*ZTL*) turbine blade factory, and the *Tsentral'nyi Kotel'no-Turbinnyi Institut* (*TsKTI*) boiler and turbine research establishment. These visits were facilitated by Mr. I. F. Maksimov and Mr V. I. Katenev of the *AO Energiya-Servis* power engineering after sales service company.

The author also wishes to thank those senior executives in Western power engineering and power generation companies, who agreed to be interviewed and participate in this project. These included Mr. M. Hangasjarvi and Mr. B. Riala of IVO International (Helsinki), Dr. J. M. Topper of CRE Group Ltd. (Cheltenham), Mr. C. Bower of Bower Energy and Environmental (Cheltenham), Mr. T. Schon of Siemens Kraftwerkunion (Erlangen), Mr. K. Harsunen of ABB Environmental Systems (Helsinki), Mr. T. Niemenen of Ahlstrom Boilers (Helsinki), Mr. J.C. McMillan of GEC-ALSTHOM European Gas Turbines (Lincoln), Mr. B. Cundill of GEC-ALSTHOM Power Engineering Systems (Whetstone), Mr. A. Thompson of GEC-ALSTHOM Combined Cycles Ltd. (Manchester), Mr. K. Bayer of GEC-

ALSTHOM EVT Energie- und Verfahrenstechnik (Stuttgart), Dr. W. MacFarlane of Rolls Royce Power Engineering (Newcastle-upon-Tyne), Mr. R. Lawrence of Parsons Power Generation Systems (Newcastle-upon-Tyne), Mr. T. Lipscomb of International Combustion Limited (Derby), Mr. A. J. Perrin, Babcock Energy (Renfrew), Mr. A. Ruffell of Babcock Energy (Crawley), Mr. M.R. Boddington of LINDEN Consulting Partnership (Lichfield), and Mr. A. Haworth of GE Power Systems (Moscow).

In addition, the author wishes to thank his academic colleagues who have provided comment and support during the research programme described in this book. Professor Julian Cooper and Mr Michael Berry of the Centre for Russian and East European Studies, University of Birmingham provided background information from Russian sources at the commencement of the project. Dr. Colin Fuller of the Centre for Hazard and Risk Management, Mr. Kenneth Turton of the Department of Mechanical Engineering and Professor Tom Weyman-Jones of the Department of Economics at Loughborough University, provided useful guidance at various stages of this research, based on their many years of experience and research in power engineering and the power generation industry; and Mr. Gerald Fish of the Business School provided information on ownership changes in the UK power engineering industry at the conclusion of the project. Dr. Harry Boer of the Faculty of Technical Management, University of Twente and Mr. Paul Walley of Warwick University Business School covered the author's teaching and administrative duties during the 1994-1995 academic year.

The information on the capacities, years of service, and fuel choices of power stations in the former Soviet Union, shown in the tables appended to Chapter 5, was calculated from data provided by Rolls Royce Power Engineering (Newcastle upon Tyne); and access to data in the Russian energy map *Toplivno-Energeticheskii Kompleks* was provided by Dr. Jonathan Stern. Data on the power stations in European Russia appended to Chapter 5 is taken from International Agency (IEA) (1994), *Electricity in European Economies in Transition*, Paris: OECD, pp. 212-23. The author acknowledges the permission of IEA for use of this material.

The author's University Department, directed consecutively by Professors D. Buchanan and J. Saunders, provided the necessary facilities to carry out this research. Initial drafts and the final manuscript of the book were prepared by Miss Lynne Atkinson and Mrs. Kathleen Gibson, and Mr. Trevor Downham drafted the diagrams shown in Chapters 3 and 4. The author wishes to acknowledge the speed, accuracy and patience of this secretarial and drafting support. Any errors in the analysis and presentation of the material provided in this book, however, remain with the author.

Author's Note

The terms 'the former USSR' and 'the former Soviet Union (FSU)' have been used throughout this book to refer to the Union of Soviet Socialist Republics (or Soviet Union), which existed before 1992 as a federal state comprised of constituent Soviet socialist republics. Many of the published data on atmospheric pollution, particularly in the base years of 1980 and 1987 chosen for subsequent international comparisons, referred to emissions from that federal state, or to the geographical region of that federal state located to the West of 60°E longitude (European USSR). It has consequently been necessary to include the whole of the former USSR as well as its European region in the research described in this book.

Following the fragmentation of the Soviet Union in December 1991, references in published data were subsequently made to emissions from each separate independent country (for example, Russian Federation, Ukraine etc.) which previously had been a former constituent Soviet socialist republic. In addition, the data referred to regions of individual countries (for example, 'European Russia located to the West of 60°E longitude').

The author has found no reference to the post-1991 'Commonwealth of Independent States (CIS)' in the international atmospheric emission published data, probably because that organization has no status as a federal state, and some states which were former Soviet socialist republics did not elect to join the Commonwealth. For those reasons, therefore, the author decided generally to refer to the 'former USSR', 'former Soviet Union (FSU)' or post-1991 individual states rather than the CIS, as individual entities; although reference is also made to the CIS where this is considered to be relevant.

To E. Flavell and A. E. Hill

1. Introduction

SCOPE OF THE BOOK

This book is divided into eight chapters which provide perspectives on atmospheric pollution in the former USSR, and the technological and commercial factors which influence the levels of these emissions. This first chapter will introduce the contents of the book, together with a short introduction to the scale of atmospheric pollution in the former USSR with particular reference to 'acid rain' emissions from the power generation industry, and an explanation of the usual parameters used to define levels of atmospheric pollution. Attention is paid to fossil fuel combustion only, as the problems of atmospheric pollution from nuclear power stations in the region have already been discussed at some length elsewhere (Lyalik and Reznikovskii, 1995; Barnaby, 1996).

The second chapter provides estimates from various published sources of the levels of emissions of oxides of sulphur (SOx) and nitrogen (NOx), paying attention to total emissions of these 'acid rain' pollutants from the former USSR, and to emissions from that region's electricity generation industry in particular. These estimates of the levels of emissions are also compared with data available from selected European countries to place the Soviet data in an international context. This discussion is then followed by descriptions in Chapters 3 and 4 of the properties of fossil fuels available in the former USSR, and of the combustion processes used in the power generation industry, with attention being drawn to the levels of SOx and NOx emissions from these various fuels and processes.

The first four chapters of this book therefore provide a contextual and technical background for a study of the capacities, ages and fuel mix of power stations in the former USSR provided in Chapter 5. The following chapter contains case studies of nine Western power engineering and generation facilities engaged in the transfer of technologies to the former USSR, and case studies of three power engineering factories and research establishments in the St. Petersburg region which are actual or potential recipients of Western technologies. The material contained in Chapters 3-6 demonstrates that the technical preconditions exist for a transfer of power engineering technologies to the former USSR, and that there is a wide range of power station capacities using various fuels having the potential to use several of these technologies to reduce emissions and improve efficiencies. Chapter 6 also shows, however, that the volume and pace of technology transfer from Western companies to organisations in the former USSR have been comparatively slow, and Chapter 7 therefore explains some of the political, economic and commercial reasons for these levels of diffusion of relevant technologies. Chapter 8 completes the book by providing comments, conclusions, and suggestions for further work arising from this research.

On a technical note, the symbols SOx and NOx have been generally used to represent 'acid rain' emissions of oxides of sulphur and nitrogen respectively. In the case of sulphur, some 98-99 per cent of SOx is usually made up of sulphur dioxide (SO_2) and reference is sometimes made explicitly to this compound in the text, especially where the published source specifically uses that symbol. NOx emissions are usually made up of two components, namely nitric oxide (NO) which accounts for some 90-95 per cent of the oxides initially released during combustion, and nitrogen dioxide (NO_2) which forms the remainder. As the nitric oxide reaches the atmosphere, however, it is almost all converted into nitrogen dioxide through a process of oxidation (Hjalmarrson, 1992, p. 14; Makansi, 1993). In some cases, though, nitrous oxide (N_2O) is also of concern as a residual pollutant from NOx control processes, in view of its activity as an extremely powerful 'greenhouse gas' which attacks the ozone layer.

The book is inter-disciplinary in content and has been written for four groups of readers. The first group are those with an interest in the problems of the global environment, who wish to know more about the scale and complexity of environmental problems in the former USSR and the practical constraints encountered when trying to reduce those problems. The second group of readers are technical and commercial personnel with a business interest in the former Soviet Union particularly, although not exclusively, in the areas of energy and environmental management, who wish to become more acquainted with the alternatives for business in that region. The third group are officials employed in government and multilateral agencies concerned with environmental and energy problems, or trade and technology transfer with the former USSR. The book is also intended to be read by academics and students with an interest in either international business or the industrial development of the former Soviet Union, as it provides information on the realities of technology and commercial activities in that region. It therefore contributes to the literature in international business and Russian studies by extending the wide range of publications on political and economic issues in that region, to the practical and urgent issues of industrial development, energy, and the environment.

The material in this book draws on a broad range of literature related to the environment in the former USSR, particularly studies carried out by atmospheric environmental specialists and geographers to arrive at estimates of the levels of atmospheric pollution in that region and the consequences for neighbouring countries. In addition, the book refers to relevant Western and Russian publications which contribute to the debate on energy and fuels policies in the former USSR as fuel selection has a direct bearing on atmospheric pollution. The major features of the book are, however, the specific attention paid to the Western and Russian industrial and technical literature related to power engineering, power generation and associated environmental issues; and case study research in three Russian power engineering establishments and nine Western power engineering companies.

The use of both Russian and Western technical sources enable comparative assessments to be made of technological developments, and the case study information provides material on the current stage of technology transfer between Western companies and industrial organisations in the former USSR. These case studies also provide information on the commercial factors affecting likely future developments in power engineering and power generation in that region.

CONTEXT OF THE BOOK

Levels of Pollution in the Former USSR

An authoritative survey on the Soviet economy prepared by the IMF, World Bank, OECD and EBRD (IMF et al., 1991, p. 1) reported that:

> sufficient information exists to conclude that many of the industrial and agricultural regions (of the former USSR) are on the verge of ecological breakdown, posing an imminent threat to the health of present and future generations.

The survey based this conclusion on the 1988 State of the (Soviet) Environment report which identified 290 ranges of severe ecological conditions covering 3.7 million square kilometres, or 16 per cent of the

USSR's total land area equivalent to an area ten times larger than unified Germany. The effects of pollution were identified as having significant deleterious effects on life expectancy, morbidity of adults and children, and genetic mutations. The causes of, and options for reduction of pollution in the former USSR, are therefore clearly important areas for future research, particularly as pollution arising from energy resources has been estimated to be approximately double that from major OECD countries (IMF et. al., 1991, p. 3), when allowance is made for their gross domestic product.

This study is specifically concerned with atmospheric pollution because of the international consequences arising from the cross-boundary transfer of pollutants. The State of the (Soviet) Environment report for 1988 indicated that about 62 million tons of primary pollutants were emitted annually into the atmosphere from Soviet industrial facilities and another 36 tons from mobile sources. These figures were considered by the IMF to underestimate the total situation by a factor of about 20 per cent, because domestic heating was omitted from the survey (IMF et. al., 1991, p. 2); and this research has consequently included a review of available data on atmospheric emissions from the former USSR (see Chapter 2). Particular attention has been paid in that chapter to SOx and NOx emissions in view of their effects on acid rain and a statement by the World Bank (1994, p. 7) that:

> the extent of the acid rain problem in Russia has not been well documented, although Russian enterprises alone discharge more SO_2 than Belgium, France, Holland, the United Kingdom and West Germany combined.

Air Pollution and Electrical Power Generation

The research described in this book has been concerned primarily with the largest industrial source of atmospheric emissions, namely electrical power generation. This sector is reported to have accounted for 25 per cent of all Soviet industrial air emissions, with oxides of sulphur (SOx) and nitrogen (NOx) comprising 65 per cent of these emissions (although the complete range of compounds within these emissions is not specified); and 43 per cent and 59 per cent respectively of total Soviet industrial emissions of SOx and NOx.[1] In other words, the electricity power generation sector has accounted for a higher proportion of the emission of these oxides when compared with its proportion of industrial air pollution as a whole, and the industry is also a significant producer of particulates and ash. In the case of sulphur dioxide emissions, it was closely followed by the non-ferrous metals industry, at 25 per cent of total industrial emissions of this pollutant.[2] In view of the effects of SOx and NOx pollutants in the production of atmospheric sulphuric acid and nitric acid precipitated as acid rain, as well as possible contributions to other atmospheric effects, the research described in this book has been mainly concerned with options for the reductions of these emissions, within the context of fuel selection and combustion processes.

Electrical Power Generation and Fuel Combustion

Electrical power is generated using the principle that an electric current can be produced by the movement of an electrical conductor within a magnetic field. Since the forces from rotary motion are easier to control in practice than those created from reciprocating motion, electricity is usually generated by rotating the turbine-driven shaft of a generator within a static surrounding magnetic field, to achieve 'turbo-generation'. The motive power for driving the turbo-generators has traditionally been steam thermal energy, converted into rotary kinetic energy in a steam turbine; and for many years coal was the primary fuel used in boilers for steam generation before its recent displacement by oil and gas. Some hydro-power has also been generated in regions where sufficient volumes and velocities of water have been available, and gas turbines were also introduced to cope with peak load variation beyond the base load capacity of conventional steam turbines (Skrotzki and Vopat, 1960; CEGB, 1986). Gas turbines have the advantage of faster start-up times than conventional steam generation sets, but their applications were originally limited to peak load generation as a consequence of the economies of scale available from the use of large capacity steam turbines for base load generation, but in recent years gas turbines have become more widely used for base load generation for reasons outlined in the following paragraph.

As coal has traditionally been the most widely used fuel for electricity power generation on a world-wide scale, international problems of atmospheric pollution have been created by the emissions of oxides of sulphur and nitrogen released during the burning of this fuel, and the use of oil shale and heavy fuel oil can also give rise to high levels of emissions of sulphur dioxide. Natural gas has recently replaced coal and oil as the major source of fuel for power generation in several countries including the former USSR (see Chapters 3 and 5) for reasons of cost, and the desirability of having a diversity of fuels available. The economic and strategic advantages to be gained from using natural gas are also enhanced by the higher efficiencies now possible from the introduction of advanced gas-fired power generation systems (see Chapter 4) and the comparative cleanliness of gas during combustion. Nuclear fission has also been developed as an alternative to the consumption of finite resources of fossil fuels which can lead to lower levels of emissions of oxides of sulphur and nitrogen, but the use of nuclear power raises other environmental concerns related to levels of radioactivity.

In spite of the development of these alternative fuels, however, coal still remains an economically viable and important fuel for power generation in certain areas of the former USSR. There are high levels of coal reserves available in the region, many of which can be used for domestic consumption, as the levels of sales of this commodity to the hard currency markets have been limited. Furthermore, significant hard currency export market opportunities continue to be seen as a priority for oil and natural gas by governments of former Soviet countries having reserves of these fuels, but current difficulties in their exploration, extraction and distribution could lead to lower quantities of these fuels being available for domestic consumption. Coal is therefore an alternative to meet some of these domestic energy shortfalls, although recent output of this fuel has also been in decline.

As some of the coals in the former USSR have a high sulphur content and others have a low calorific value (see Chapter 3), however, attention to the problems of coal-fired atmospheric pollution will remain high on the ecological agenda, if that fuel is awarded a higher priority status. Any favourable economic considerations influencing the selection of coal and other fossil fuels may also be reinforced by even greater ecological concerns over the possible discharge of atmospheric radiation from Soviet-designed nuclear power stations, particularly when these generation facilities are operated intensively.

In the present changing political and economic circumstances in the former USSR, though, the economic viability of some coal deposits may continue to change. For example, there are difficulties in continued working of narrower seams, and the product has a high sulphur content in some of the traditional mining areas of European Russia; and the long distances between Siberian coal deposits and centres of industry and population in European Russia and the Urals can also lead to high transport costs between these regions. In the previous Soviet planned economy, most of these costs were covered by subsidies as prices of key commodities were centrally approved; but many of these forms of government support and control continue to be removed during the process of transition to a post-Soviet variant of the market economy. It is difficult at this point in time to predict the effects of subsidy removal on energy mix in the former USSR, particularly if subsidies and price controls are removed from all sources of energy in addition to coal.

Fuel selection policies will clearly have a direct effect on the levels of SOx and NOx emissions and a major consequent influence on attempts to reduce the dimensions of atmospheric problems in the former USSR. Information on fuel selection and combustion processes is therefore provided in Chapters 3 and 4.

Atmospheric Emissions from Thermal Power Stations

The emissions from thermal power stations during the combustion process can be defined in three main ways, and the control and reduction of atmospheric pollution are usually achieved by the setting of targets relating to a specific parameter or set of parameters. The three major categories are total emissions of a particular pollutant for a defined time interval; the quantity of pollutant per unit of energy generated; or the quantity of pollutant per unit volume of emission.

Each of these categories are described in the following paragraphs using data available from Boehmer-Christiansen and Skea (1991, pp. 234-50). Further information on the capacity and output of the power generation industry in the former USSR is provided later in this book (Chapter 5), as these factors combine with fuel properties and combustion processes to influence each of the categories described below.

Total emissions of a particular pollutant (for example, NOx or SOx) for a defined time interval (for example, per calendar year)

Total emissions are sometimes referred to as 'pollution budgets', and targets may be set for their reduction by a specified date. These targets are usually defined as 'national budgets' for individual countries, as in the case of the United Nations' Economic Commission for Europe (ECE) Long Range Transboundary Air Pollution (LRTAP) Protocols and the European Commission (EC) Large Combustion Plant (LCP) Directive, and each country will then disaggregate these national budgets amongst individual power generating organizations.

The 1985 Helsinki Protocol committed the signatory ECE countries (including the former USSR) to reduce either SOx emissions or transboundary transmissions of SOx pollution by 30 per cent of the 1980 level, by the year 1993. A second SOx protocol was signed in Oslo in June 1994 relating to targets (or 'national budgets') for the years 1995, 2000, 2005 and 2010. An ECE protocol which specified a freeze in NOx emissions from 1994 onwards, which were set at the 1987 level, was signed in Sofia in 1988. The former USSR was also a signatory to this NOx protocol, and the reduction of emissions from the region to meet the Helsinki and Sofia Protocols is discussed in more detail in Chapter 7.

In the case of the EC LCP Directive, total emissions of SOx and NOx from large combustion plants (that is, greater than 50 MW capacity) in each member state were to be reduced to defined levels in specific years, calculated from 1980 emissions as a base level. Although the defined national emission 'bubbles' varied from country to country, they were based on 60 per cent reductions of SOx and 40 per cent reductions of NOx, by 1995. The USSR was not bound by the requirements of this directive as it was not a member of the European Union, but similar future targets might well be placed on the European states which were previously constituents of the former USSR (as well as other former socialist countries of Eastern Europe) as these republics continue to move towards closer economic links with the West European trading community.

The quantity of pollutants emitted per unit of heat energy or electrical energy generated
Levels of atmospheric emissions are influenced by the calorific values, chemical contents, and combustion properties of the fuels burnt; and the efficiency and control of the combustion process. Reference is frequently made to these parameters by different researchers both in the West and the former USSR but comparison is often difficult because of the use of both metric and imperial units, and the application of different procedures for measurement and test. They can be quite useful, nevertheless, as they provide an indicator of the effects of fuel characteristics, combustion processes and plant efficiency, on SOx and NOx emissions.

The quantity or weight of emissions contained in a defined volume of air emitted from the power station into the atmosphere
The parameter chosen to define these emissions is usually either 'parts per million' (ppm) of a pollutant in emission gases, or milligrams (mg) of pollutant (for example, SOx or NOx) contained in a cubic metre of air (Nm^3), 'normalised' to defined conditions of pressure and temperature. This volumetric parameter is frequently chosen in preference to those defined in the previous sub-section, as it is a fairly straightforward measure of pollutant emission to the atmosphere, irrespective of the fuel; and provides a focus for specific combustion processes and post-combustion treatments, used for particular fuels. The EC Large Combustion Plant Directive approved in 1988 provides limits for emissions of SOx and NOx defined by mg/Nm^3 parameters, for new power stations having a generating capacity of more

than 50MW. For new power stations of greater than 500MW capacity, the SOx emission limits were set at 400 mg/Nm^3 for solid and liquid fuels, and 35 mg/Nm^3 for gaseous fuels. For NOx emissions, the corresponding limits were 650 mg/Nm^3 for solid fuels, 450 mg/Nm^3 for liquid fuels, and 350 mg/Nm^3 for gases. In the case of power stations lower than 500MW capacity, SOx emission limits increased on a linear scale from 400-2,000 mg/Nm^3 for solid fuels as capacity decreased from 500MW to 300MW. For liquid fuels, emission limits increased on a linear scale from 400-1700 mg/Nm^3 as capacity decreased from 500 to 300 MW capacity. For those plants burning indigenous solid fuel, where the SOx emission limits would entail the use of excessively expensive technology, desulphurization rates were to be reduced by 90 per cent for plant sizes above 500MW, and on a sliding scale from 90 per cent to 40 per cent as plant capacity decreased from 500MW to 167MW. Several European governments have specified more stringent requirements than those specified in the EC Large Combustion Plant Directive (for example, limits of 200mg/Nm^3 for SOx emissions from West German power stations). In the case of existing power stations, however, the required national reductions in total SOx and NOx emissions referred to above, can be disaggregated to reductions in volumetric specific emissions in individual power stations.

The USSR had published a series of specifications and standards on emissions before the fragmentation of that that federal state in December 1991, and the Russian Federation has subsequently published other regulations (see Chapter 2). Many of the parameters in these documents conform to, or are better than, the requirements of the EC Large Combustion Plant Directive but the data presented in Chapter 4 reveals that emissions from combustion processes in many power stations of the former USSR exceed those of West European normative documents.

NOTES

1. IMF, et al. (1991), pp. 2, 3. These figures for the power generation industry from IMF et. al (1991, pp. 2, 3) are also confirmed from another source, namely Gushcha (1993) who quotes 24 per cent of total industrial emission, 42 per cent of sulphur anhydride (SOx) and 60 per cent of NOx.

2. The 25 per cent figure for SOx emissions from the non-ferrous metals industry was explicitly quoted in IMF (1991, p. 3) apparently using data from the 1988 State of the (Soviet) Environment report, and it has consequently been assumed that this industry is the second largest contributor of this pollutant. Ferrous metallurgy is also a heavy polluter (quoted as 17

per cent of all atmospheric emissions in the IMF report), but a separate SOx emissions figure is not given for that industry.

REFERENCES

Barnaby, F. (1996), 'Nuclear Accidents Waiting to Happen', *The World Today,* Volume 52, Number 4 (April 1996) pp. 93-6.

Boehmer-Christiansen, S. and Skea, J. (1991), *Acid Politics : Environmental and Energy Politics in Britain and Germany,* London: Belhaven.

Central Electricity Generating board (CEGB) (1986), *Advances in Power Station Construction,* Oxford: Pergamon.

Gushcha, V.I. (1993), 'Puti resheniya ekologicheskikh problem na ugol'nykh elektrostantsiyakh Rossii' (Ways to solve ecological problems at Russian coal-fired electrical power plants). Paper presented at the Symposium on New Coal Utilization Technologies, Helsinki, 10-13 May 1993 (Economic Commission for Europe, Committee on Energy, Working Party on Coal).

Hjalmarrson, A.K. (1992), *Interactions in emissions control for coal-fired power plants,* London: IEA Coal Research.

International Monetary Fund (IMF), the World Bank, the Organization for Economic Cooperation and Development (OECD), the European Bank for Reconstruction and Development (EBRD) (1991), A *Study of the Soviet Economy, Volume 3,* Paris: OECD.

Lyalik, G.N. and Reznikovskii, A.Sh. (1995), *Elektroenergetika i priroda,* Moscow: Energoizdat.

Makansi, J. (1993), 'Reducing NOx Emissions from Today's Power Plants'. *Power,* May 1993, pp. 11-31.

Skrotzki, B.G.A. and Vopat, W.A. (1960), *Power Station Engineering and Economy,* New York: McGraw-Hill.

World Bank (1994), *Staff Appraisal Report, Russian Federation, Environmental Project,* Report No. 12838-RU, , Washington, DC: World Bank.

2. Atmospheric Pollution in the Former USSR - Dimensions and Context

ATMOSPHERIC EMISSIONS FROM THE FORMER USSR

Introduction

As discussed in more detail in a subsequent chapter of this book (see Chapter 7), atmospheric emissions during 1980 for SOx and 1987 for NOx, were used as international benchmarks for target reductions in future years. It is therefore important to arrive at realistic estimates of those 'acid rain' emissions at that time for both the former USSR and the Russian Federation, as bases for international comparison and objectives for reduction during subsequent years.

Data on the emissions of oxides of sulphur (SOx) and nitrogen (NOx) from the former USSR during the early 1980s are available from one Russian and nine Western sources (Gushcha, 1993; Dovland, 1987; Pacyna et al., 1991; Pryde, 1991; Ramus 1991; Veldt, 1991; Mylona, 1993; Tuovinen et al., 1993; ECE, 1994; Tuovinen et al., 1994), although it is not always clear in the Russian source (Gushcha, 1993) whether the data is based on official government data, and whether that in turn is based on records or estimates. It is more likely to be based on estimates, however, as Western visitors to power stations and research establishments in the former USSR have informed the author that instrumentation for such recording is so rare as to be almost non-existent. The analysis of these data is made difficult by the fact that sources refer variously to either total emissions, stationary emissions or industrial emissions for different years, and it is not altogether clear whether 'transport' is counted as an industry in these reports. These sources also variously refer to the geographical regions defined either by the former USSR in total, the Russian Federation, the European regions of the former USSR, or European Russia. In the latter definitions referring to European Russia and the European regions of the former USSR, the 60°E longitude line is often taken as the easternmost limit of Europe, although this definition includes only part of the Ural Mountains which is an important industrial region. The boundary on the EMEP map used by the Economic Commission for Europe appears to be marginally west of the Urals in some parts, however, and may therefore exclude some of the industrial cities in that region. Veldt (1991) on the other hand, specifically included the Ural Mountains in his study of SOx and NOx emissions, and thereby included the industrial cities of Chelyabinsk and Ekaterinburg (formerly Sverdlovsk) which are located to the east of 60°E longitude.

Aggregate SOx Emissions

Ramus (1991) provides an estimate of 16.2M tonnes of SOx emissions annually from the former USSR during 1980/82 (see Table 2.1), and Gushcha provides a figure of 10.5 million tonnes of SOx in 1980 from the Russian Federation, reducing to 9.1M tonnes in 1985, and to 6.9M tonnes in 1990 (see Table 2.2). If it is assumed that the mix of fuels and combustion conditions were approximately the same throughout all of the republics of the former USSR and that atmospheric emissions were approximately proportional to the volume of power generated[1] then the SOx emissions from the former USSR can be estimated as 16.9M tonnes in 1980 falling to 14.6M tonnes in 1985, 12.2M tonnes in 1988, and 11.7M tonnes in 1989 (see Table 2.2), although these estimates can be criticised for being based on total power produced, rather than power produced in thermal power stations.

More detailed information on Russian atmospheric pollution is available from the EMEP (Co-operative Programme for Monitoring and Evaluation of Long Range Transmission of Air Pollutants in Europe) surveys carried out by the Meteorological Synthesizing Centre-West (MSC-W) of the Norwegian Meteorological Institute (Tuovinen et al., 1994), which provide data on emissions of sulphur dioxides from all of Europe, including the European regions of the former USSR (see Table 2.3). These data confirm that the European region of Russia has been by far the single largest contributor to SOx on the European continent; and Ukraine was only exceeded by UK, Poland and the GDR in 1980.

The data also show, however, that the levels of European Russian sulphur dioxide emissions were reduced by some 38 per cent between 1980 and 1992, and the significance of these changes will become apparent in a subsequent section of Chapter 7 which discusses international long range protocols for transboundary air pollution. A comparison of the data from Tables 2.2 and 2.3 also suggests that European Russia accounted for some

Table 2.1 Sulphur Dioxide and Nitrogen Oxide Emissions from Selected European Countries (Thousands of Tonnes per Year: Data shown for early 1980s)

Country	SO₂ Emissions	NOx Emissions
Austria	440	n.a.
Belgium	810	290
Denmark	456	181
Federal Republic of Germany	3,630	1,895
France	3,600	1,297
Ireland	172	60
Italy	3,800	n.a.
Netherlands	480	401
Norway	150	102
Sweden	550	247
Switzerland	124	n.a.
United Kingdom	5,340	1,888
Czechoslovakia	3,000	n.a.
German Democratic Republic	4,000	n.a.
Hungary	1,632	n.a.
Poland	4,300	n.a.
USSR	16,200	n.a.

Source: Ramus (1991), pp. 8, 9 citing EMEP calculations for October 1980-April 1981 and March-October 1982 (EMEP/MSC-W) and *Strategies and Policies for the Abatement of Air Pollution* (ECE/EB/AIR/rev. 1/20-2-85) for sulphur emissions. Nitrogen data is taken from Derwent (1985).

Table 2.2 Data on Atmospheric Emissions and Electricity Generation in Russia and the Former USSR

Year	Statistical Data (1980-90)					Energy Programme (1995-2010)			
	1980	1985	1988	1989	1990	1995	2000	2005	2010
Total (for Russia)[1] M tonnes including	19.3	17.9	15.3	14.6	14.0	13.5	13.0	12.5	11.8
SOx	10.5	9.1	7.6	7.3	6.9	7.3	6.9	6.5	6.0
NOx	2.3	2.4	2.7	2.7	2.8	2.8	2.8	2.8	2.8
Estimated SOx emissions from former USSR[2]	16.9	14.6	12.2	11.7					

Notes:
1. Data taken from Gushcha (1993), p. 3.
2. Calculated from proportionate increases of cited figures for Russian SOx emissions, according to the annual ratio of Soviet power generation to Russian power generation (that is for 1980 and 1985 respectively: $16.9 = 10.5 \times 1294/805$; $14.6 = 9.1 \times 1544/962$ etc.) from the power generation data shown below:

Year	1980	1985	1988	1989
Electricity output (USSR Total TWh)	1294	1544	1705	1722
Electricity Output (Russia TWh)	805	962	1066	1077

Source: Goskomstat (1990),. p. 375.

Table 2.3 SOx Emissions from European Regions of the Former USSR (1980-1992)

Emissions of Sulphur Dioxides (1000 tonnes of SO_2 per annum)

Year	Belarus	Russia	Ukraine	Estonia	Latvia	Lithuania	Moldova	Total
1980	740	7161	3850	275	90	136	91	12343
1985	690	6191	3464	240	90	136	91	10902
1986	690	5707	3392	240	90	136	91	10346
1987	690	5622	3264	240	90	136	91	10133
1988	638	5145	3211	240	90	136	91	9551
1989	596	4677	3073	240	90	136	91	8093
1990	596	4460	2782	240	82	136	91	8387
1991	596	4392	2538	235	82	136	91	8070
1992	596	4400	2376	180	82	136	91	7861

Note: SO_2 emissions in 1980 for UK, the German Democratic Republic, Poland and Ukraine are given as 4900, 4260, 4100 and 3850 thousand tonnes, respectively.

Source for EMEP data presented in Table 2.3: Tuovinen et al. (1994), pp. 24, 26.

64-68 per cent of total SO_2 emissions from the Russian Federation[2] between 1980 and 1989. It is important to bear in mind, however, that the data provided by the sources cited in Tables 2.1 and 2.3 are based on official government estimates, as the authors of the 1994 EMEP report are explicit that their estimates are based on officially reported data to the United Nations Economic Commission for Europe (ECE) from the governments of Belarus, Russian Federation, Ukraine, and the former Soviet Union. Similar data were probably also used for the EMEP report cited by Ramus, and official government estimates were probably used for the emissions data in Table 2.2 in view of the consistencies in the proportion of total Soviet emissions accounted for by the Russian Federation. The authors of the 1994 EMEP publication, however, also report significant differences between the officially submitted data and other estimates, particularly those by Veldt (1991), Pacyna et al. (1991), Tuovinen et al. (1993) and Mylona (1993), as discussed in the subsequent paragraphs of this section.

As mentioned in Chapter 1, IMF et al. (1991, p. 2) pointed out that emissions from domestic heating were omitted from the 1989 State of the (Soviet) Environment survey in which industrial SOx emissions were reported as 17.6-18M tonnes for 1988.[3] This omission may have led to an underestimate of total emissions by some 20 per cent and Tuovinen et al. (1993) also point out that emissions from small boilers and furnaces have been neglected in the official Soviet government inventories of SOx and NOx. In practice, however, the effects of these omissions could be similar, as the American use of the term 'furnace' often refers to space heating combustion equipment, as distinct from the British use of the term which usually refers to industrial processing facilities for the heating of materials.

If the cited underestimate of 20 per cent applies equally to all emissions including SOx, then the total SOx emissions from the former USSR in 1980 can be estimated to have been some 20.3M tonnes rather than the 16.2M tonnes quoted in Table 2.1. Evidence available from Ryaboshapko (1990) cited by Tuovinen et al. (1993) suggests, however, that the underestimate of 20 per cent cited by IMF may have been too low when applied to SOx emissions. For 1990, Ryaboshapko cites an official Soviet (*Goskomstat*) emission level for SOx of 16M tonnes, whereas his own research including inventories for small boilers and furnaces suggests a level closer to 25M tonnes. If this underestimate of 36 per cent also applies to the 1980 figure of 16.2M tonnes for the USSR cited by Ramus, it is possible that the actual 1980 figure should have been 25.3M tonnes.

There is also evidence available from Veldt (1991) using estimates of fuel consumption in stationary emission sources and vehicle mix for mobile emission sources, that the levels of air pollutants in the European region of the former USSR (that is West of the 60°E longitude line, plus all of the Ural Mountains) are far greater than those that would be expected from official reports. Veldt provides a total figure of 19.4M tonnes for SOx emissions 1985 (see Table 2.4), which is 76 per cent higher than the ECE (1987) reported levels of 11.1M tonnes of SOx cited by Dovland, and a major difference even when allowance is made for the influence of the Urals region. If Veldt's assessment of underestimation is correct, and applies to the former USSR in 1980 as well as the European regions of that federal union in 1985, then estimated SOx emissions of 16.2M tonnes for the former USSR in 1980 should be likewise increased by 76 per cent to 28.5M tonnes, although there may be some double counting of emissions for part of the Urals region in this estimate. Dovland's estimate corresponds closely to the EMEP data for Belarus, European Russia and Ukraine cited for 1985 in Table 2.3 (totals of 10.9M tonnes of SOx) although consistency between the ECE and EMEP reports is to be expected in view of their common data sources of official government statistics. Any variations are probably due to modifications to the data between the 1987 publication by Dovland of the 1985, data and the 1994 publication of the most recent data by Tuovinen et al. (1994). It is necessary to note, however, that according to Tuovinen et al., Mylona's estimates do not significantly differ from official Soviet SOx emission data for 1980, although there are differences in the levels of reduction in subsequent years.

To conclude, therefore it is difficult to say which of the estimates for SOx emissions is correct (see Table 2.5) as the smaller values are based on government reports, whilst the larger values are based on calculations allowing for domestic combustion and the properties of available fuels. For the purposes of this research therefore, a range of emissions has been selected for comparisons of emissions from other countries, and also to assess the levels of reduction in subsequent years. The present author is inclined to favour the medium and higher values shown in Table 2.5, as these values account for emissions from domestic and other non-industrial small boilers. Even if official Soviet and Russian data are used, however, the former USSR still emerges as a major contributor to international SOx emissions and the major emitter of SOx on the European continent, with levels of SOx emissions approximating to the levels of the US (some 23.8M tons in 1980 according to ECE, 1994) rather than those from any other European country. Although the power generation sector has been a major source of a significant proportion of these emissions, and is the major focus of this research, non-industrial boilers have also been important contributors of SOx emissions and candidates for the introduction of SOx control technology.

Table 2.4 Sources of SOx Emissions in the European Regions of the Former USSR (1985: Million Tonnes)

Sources of Emissions	SOx	
Utility and Industrial Boilers	6.4	(33%)
Non-industrial Small Boilers	11.4	(58%)
Refineries and Chemical Processes	1.4	(7%)
Subtotal: Stationary Sources	19.2	(99%)
Transportation	0.2	(1%)
Total Emissions	19.4	(100%)

Source: Adapted from Veldt (1991), Table 10. The differences between the quoted and summated sources are due to rounding errors.

Aggregate NOx Emissions

An estimate of 4.5M tonnes for the levels of NOx emissions from the former USSR in 1987 is provided by Pryde (1991) based on the figure cited in the official Soviet statistical handbook for that year (*Narodnoe khozyaistvo SSSR v 1987g*). A figure of 4.5M tonnes is also provided by Pryde for industrial emissions in 1988 cited from the *Goskompriroda* (1989) report on the state of the Soviet environment, whilst the official Soviet statistical handbook for that year provides a figure of 5M tonnes for NOx emissions in a footnote to a table which provides data on emissions from stationary sources.

According to research carried out by Veldt (1991) and Pacyna et al. (1991), on emissions from the European regions of the former USSR, however, these official data may underestimate the total levels of NOx emissions from the former USSR. In 1987, for example, the ECE provided a figure of 2.9M tonnes for NOx emissions from the European regions of the former USSR (cited by Dovland, 1987), whilst the summation of emissions data for each constituent republic of the European region of the former USSR provided a total of some 3.5M tonnes for 1985 (see Table 2.6). These official figures,however, are lower than the figure of 4.4 and 4.2M tonnes provided

Table 2.5 Estimates of SOx Emissions (1980: Million Tonnes)

Region	Official Data (M tonnes)	Official Data Corrected for Domestic Boilers (official data increased by a factor of 56% see Tuovinen et al., 1991)	Official Data Corrected for Non-industrial Small Boilers (official data increased by a factor of 76%, (see Veldt, 1991)
	(Column 1)	(Column 2)	(Column 3)
European Russia	7.2[1]	11.2	12.6
Russian Federation	10.5[2]	16.4	18.5
European regions of former USSR	12.3[3]	19.3	21.7
Former USSR	16.2[4]	25.3	28.5

Notes:
1. See Table 2.3
2. See Table 2.2
3. See Table 2.3
4. See Table 2.1

for stationary sources by Veldt and Pacyna et al. (see Tables 2.7 and 2.8) and appear to take no account of NOx emissions from transportation even though that sector may have been viewed as an 'industry' in the *Goskompriroda* estimate.

Veldt and Pacyna et al. provide an estimate of 7.2M tonnes and 7.1M tonnes respectively for the European region of the former USSR (see Tables 2.7 and 2.8), suggesting an underestimate of some 60 per cent in the official figures cited by Dovland as a consequence of the omission of transportation emissions from the usually cited Soviet data. If this correction factor for underestimation is applied to the 4.5M figure quoted for the former USSR in 1987, then this latter figure is increased to 11.2M tonnes, with other values for European Russia, the Russian Federation and the European region of the former USSR as shown in Table 2.9. When the various estimates for the former USSR are compared with emissions from other European countries, as shown in the footnote to Table 2.6, it is apparent that the European region of the former USSR has been the single largest contributor to NOx emissions on the European continent. Although a significant proportion of these emissions has been from the power generation sector, which is the major focus of this research, it is also important to note that the transportation sector has also been a major source of NOx emissions and therefore an important candidate for the introduction of NOx control technologies. It is important to note, however, that even though the former USSR has been the largest single contributor of NOx emissions in Europe, other international comparative data reveal that those emissions have been far lower than those of the US (18.7M tonnes in 1987, according to ECE, 1994) even when allowance is made for an underestimate of the official Soviet figure.

Local Levels of SOx and NOx Emissions

In view of the statements related to high levels of pollution in the IMF report cited at the commencement of this book, it is necessary to consider data available on local levels of pollution, as well as the national levels discussed in the previous section of this chapter, in order to investigate local variations from the national average. In addition to the above-cited national data, the former USSR also provided official statistics on the levels of SOx and NOx emissions for certain cities and per capita data have been calculated using the population figures given in Table 2.10 as a denominator. The cities have then been ranked in terms of their total SOx and NOx emissions, using a running total of population equivalent to 10 per cent of the grand total of the Soviet population as a cut-off point (see Table 2.11).

Table 2.6 NOx Emissions from European Regions of the Former USSR (1980-1992)

Emissions of Nitrogen oxides (1000 tonnes of NO_2 per annum)

Year	Estonia	Latvia	Lithuania	Belarus	Russia	Ukraine	Moldova	Total
1980	66	54	56	244	1734	1059	35	3248
1985	66	54	56	220	1903	1059	35	3393
1986	66	54	56	258	1871	1112	35	3452
1987	66	54	56	287	2353	1095	35	3946
1988	66	54	56	262	2358	1090	35	3921
1989	66	54	56	263	2553	1065	35	4092
1990	66	54	56	263	2675	1097	35	4246
1991	64	54	56	263	2571	990	35	4033
1992	64	54	56	263	2326	830	35	3628

Note: NO_2 emissions in 1987 were 3050 thousand tonnes per year for the German Federal Republic, 2603 thousand tonnes per year for UK, and 1700 thousand tonnes per year for Italy.

Source: Tuovinen et al. (1994), pp. 24, 26.

Table 2.7 Sources of NOx Emissions in the European Region of the Former USSR (1985: Million Tonnes)

Sources of Emissions	NOx
Utility and Industrial Boilers	1.2 (17%)
Non-industrial Small Boilers	2.7 (38%)
Refineries and Chemical Processes	0.5 (7%)
Subtotal: Stationary Sources	4.4 (62%)
Transportation	2.8 (39%)
Total Emissions	7.2 (100%)

Source: Adapted from Veldt (1991), Table 10. The differences between the quoted and summated percentages are due to rounding errors.

Table 2.8 Anthropogenic NOx Emissions in the European Region of the Former USSR (1985: Million Tonnes)

Source of NOx emissions	Volume of NOx emissions
Stationary Sources	4.2 (59%)
Mobile Sources	2.9 (41%)
Total	7.1 (100%)

Source: Adapted from Pacyna et al. (1991), Table 7

Table 2.9 Estimates of NOx Emissions (1987: Million Tonnes)

Region	Official Figures	Official Figures increased by 148% (see Veldt 1991 and Pacyna et al. 1991)[5]
(Column 1)	(Column 2)	(Column 3)
European Russia	2.4[1]	5.8
Russian Federation	2.6[2]	6.4
European region of former USSR	3.9[3]	9.7
Former USSR	4.5[4]	11.2

Notes:
1. See Table 2.7.
2. Estimated from 1988 figure in Table 2.2.
3. See Table 2.7.
4. See Pryde (1991).
5. That is, to account for apparent underestimate by 60% as a result of omitting data on transport from the official figures in Column 2.

From these data, it can be seen that certain cities (for example, Moscow and Leningrad) have had very high levels of SOx emissions, which are partly to be expected in view of the quantity of stationary power and heating sources required to serve their large populations. In addition, however, there are also cities with moderate levels of population (for example, Ufa, Novokuznets, Magnitogorsk, Mariupol and Ust-Kamenogorsk) which have had particularly high levels of SOx and NOx emissions as a result of their high levels of industrial activity (particularly of metallurgy) or their use of particular types of coal (for example, high sulphur Ukrainian or Kazakh coals). The influence of these factors is highlighted by the ranking in Table 2.12 which shows very high levels of per capita SOx and NOx emissions from certain cities of moderate size, but particularly Ust-Kamenogorsk, Magnitogorsk, Novokuznets and Mariupol.

Although the above-cited data is useful to obtain an impression of the levels of SOx and NOx emissions from particular urban locations, it is important to note that although these cities accounted for 14.5 per cent of the total Soviet population in 1987, they only accounted for some 7.1 per cent of the total SOx emissions and some 12.7 per cent of the total NOx emissions. This suggests that although there are several cities included with very high levels of emissions there are also several other important industrial cities which are not included such as Norilsk, Zapolyarnyi, Khabarovsk (Bezuglaya 1995), Sverdlovsk and Gor'kii. Alternatively, there may be many stationary sources established in locations of smaller populations than those shown in the list which emit high levels of SOx and NOx as a consequence of proportionately higher levels of pollutant emission per unit of power generated from smaller (and probably less efficient) installations. In spite of these omissions from the data shown in Table 2.10, however, it is evident from the data shown in Tables 2.11 and 2.12 that there are several cities in the former USSR with particularly high levels of SOx and NOx emissions; and especially high levels of SOx and NOx emissions per capita of population. These cities are clearly prime candidates for the installation of pollutant control equipment.

Power Station SOx and NOx Emissions

In addition to the national and local effects of SOx and NOx pollution referred to in the previous sections of this chapter, it is also important to take account of specific industrial sources of atmospheric emissions, as these have a consequential local effect. As electric power generation is the single largest contributor to industrial SOx and NOx emissions, particular attention has been paid to that sector in this book. Turning initially to the official emission figure of 16.2M tonnes of SOx in 1980, and the view of most of the above cited researchers that official emission data for SOx refer to stationary industrial sources only, combined with the often repeated statistic that the power generation sector accounts for 43 per cent of industrial SOx emissions, then power generation SOx emissions can be estimated to have been some 7.0M tonnes in 1980. This figure is lower, however, than if Veldt's data is used based on the estimate of the proportion of total SOx emissions accounted for by utilities and industrial boilers (that is, 33 per cent in 1985, see Table 2.4). If this proportion is applied to the estimate of total SOx emissions in 1980 (28.5M tonnes), the figure emerges of 9.4M tonnes of SOx emissions from power stations during that year. This estimate would be a maximum, however, as some industrial boilers will not be used for the generation of electricity, but for the generation of process heat; and Veldt's data for European regions of the USSR included all of the Ural Mountain region before scaling up for the USSR as a whole.

Novozhilov and Evdokimova (1995) provide a figure of 5.9M tonnes of SOx emissions from Russian thermal power stations in 1980, which can be extended to an estimate of 9.5M tonnes for Soviet power stations during that year, if it is assumed that the emissions vary in proportion to the power produced and the mix of fuels used in power generation was similar across the former USSR. Lyalik and Reznikovskii (1995, p. 118) provide a slightly higher figure of 10.6M tonnes for SOx emissions from Soviet power stations in 1980, and a range of emissions from 7.0M tonnes to 10.6M tonnes is consequently provided in Table 2.13.

Similarly, if the official estimate for NOx emissions in 1987 is assumed to have been some 4.5M tonnes, and the view of most of the above cited researchers accepted that official emission data for NOx refers to stationary industrial sources only, combined with the often-repeated statistic that the power generation sector accounts for some 59 per cent of industrial NOx emissions, then power station NOx emissions from the former USSR can be assumed to have been some 2.7M tonnes in 1987. This figure lies between the 1985 level of 2.41M tonnes and the 1990 level of 2.7M tonnes quoted by Lyalik and Reznikovskii (1995, p. 118) for Soviet power stations, and has therefore been quoted in Table 2.13. Applying the same factor of 59 per cent to the NOx emissions data shown for European Russia, the Russian Federation and the European region of the former USSR, then the estimates for power station NOx emissions would have been 1.4M tonnes, 1.5M tonnes, and 2.3M tonnes respectively for those regions.

Table 2.10 SOx and NOx Pollutants from Stationary Sources for Selected Soviet Cities (1987)

City	Population (thousands)	SO_2 emissions (tonnes)	SO_2 per capita (tonnes per thousand)	NOx emissions (tonnes)	NOx emissions per capita (tonnes per thousand)
Alma Ata	1,108	15,700	14	3,800	3.4
Arkhangelsk	416	38,400	92	7,500	18.0
Ashkabad	382	400	1	200	0.5
Baku	1,741	20,300	12	16,500	9.5
Bratsk	249	22,000	88	4,700	18.9
Chelyabinsk	1,119	60,700	54	29,400	26.3
Dzhambul	315	51,700	164	13,900	44.1
Donetsk	1,090	33,700	31	7,200	6.6
Dushanbe	582	8,300	14	4,300	7.4
Frunze	632	42,100	67	7,900	12.5
Irkutsk	609	26,400	43	8,100	13.30
Kiev	2,544	39,100	15	22,100	8.7
Kemerovo	520	23,900	46	29,700	57.1
Kishinev	663	18,800	28	5,400	8.1
Krasnoyarsk	899	40,200	45	12,600	14.0

Leningrad	4,948	85,000	17	45,500	9.2
Mariupol	529	55,000	104	30,100	56.9
Magnitogorsk	430	81,700	190	34,400	80.0
Minsk	1,543	29,500	19	17,300	11.2
Moscow	8,815	113,800	13	117,800	13.4
Mogilev	359	70,600	197	7,500	20.9
Novokuznetsk	589	90,900	154	36,700	62.3
Odessa	1,141	20,100	18	6,700	5.9
Riga	900	9,900	11	2,800	3.1
Tallin	478	19,700	41	4,200	8.8
Tashkent	2,124	4,000	2	4,800	2.2
Tbilisi	1,194	4,000	3	3,400	2.9
Ust-Kamenogorsk	321	73,300	228	8,100	25.2
Ufa	1,092	101,800	93	27,100	24.8
Volgograd	988	46,400	47	19,600	19.8
Vilnius	566	22,000	39	4,900	8.7
Yerevan	1,168	19,600	17	11,700	10.0
Zaporozhye	875	27,400	31	14,600	16.7
Sub-total for listed cities	40,929	1,316,400	32	570,000	14
Reported total for USSR	281,689	18,600,000	66	4,500,000	16

Note: Population, SO$_2$ emissions and NOx emissions taken from Pryde (1991), pp. 34,35 (Pryde cites *Narodnoe khozyaistvo SSSR v 1987g*, p. 573)

Table 2.11 SOx and NOx Pollutants from Stationary Sources for Selected Soviet Cities (1987: Ranked in Order of Levels of Emissions)

City	Population (thousands)	SOx emissions (tonnes)	City	Population (thousands)	NOx emissions (tonnes)
Moscow	8,815	113,800	Moscow	8,815	117,800
Ufa	1,092	101,800	Leningrad	4,948	45,500
Novokuznetsk	589	90,900	Novokuznetsk	589	36,700
Leningrad	4,948	85,000	Magnitogorsk	430	34,400
Magnitogorsk	430	81,700	Mariupol	529	30,100
Ust-Kamenogorsk	321	73,300	Kemerovo	520	29,700
Mogilev	359	70,600	Chelyabinsk	1,119	29,400
Chelyabinsk	1,119	60,700	Ufa	1,092	27,100
Mariupol	529	55,000	Kiev	2,544	22,100
Dzhambul	315	51,700	Volgograd	988	19,600
Volgograd	988	46,000	Minsk	1,543	17,300
Frunze	632	42,100	Baku	1,741	16,500
Krasnoyarsk	899	40,200	Zaporozhye	875	14,600
Kiev	2,544	39,100	Dzhambul	315	13,900
Arkhangelsk	416	38,400	Krasnoyarsk	899	12,600
Donetsk	1,090	33,700	Yerevan	1,168	11,700

Minsk	1,543	29,500	Irkutsk	609	8,100
Zaporozhye	875	27,400	Ust-Kamenogorsk	321	8,100
Irkutsk	609	26,400			
Kemerovo	520	23,900			
Total for listed cities	28,683	1,131,200		29,047	495,200

Source: Table 2.10

Table 2.12 SOx and NOx Emissions per Capita for Selected Soviet Cities (Ranked in Order of Levels of Emissions)

City	SOx emissions (tonnes per thousand of population)	City	NOx emissions (tonnes per thousand of population)
Ust-Kamenogorsk	228	Magnitogorsk	80
Mogilev	197	Novokuznetsk	62
Magnitogorsk	190	Kemerovo	57
Dzhambul	164	Mariupol	57
Novokuznetsk	154	Dzhambul	44
Mariupol	104	Chelyabinsk	26
Ufa	93	Ust-Kamenogorsk	25
Arkhangelsk	92	Ufa	25
Bratsk	88	Mogilev	21
Frunze	67	Volgograd	20
Chelyabinsk	54	Bratsk	19
Volgograd	47	Arkhangelsk	18
Kemerovo	46	Zaporozhye	17
Krasnoyarsk	45	Krasnoyarsk	14
Irkutsk	43	Moscow	13
Tallinn	41	Irkutsk	13
Vilnius	39	Frunze	13

Source: Table 2.10

These estimates, however, are higher than those obtained from multiplying Veldt's estimated proportion of 17 per cent for utility and industrial boilers (see Table 2.7) by the estimate of total emissions. If such a calculation was carried out using the 1987 estimate of 11.2M tonnes, an estimate of 1.9M tonnes would be obtained for power station NOx emissions from the former USSR.

Table 2.13 also provides comparative data on estimates of levels of emissions of SOx and NOx from power stations for UK, Germany and the former USSR during 1980 and 1987 respectively. From these data, it appears that during those years, total emissions from electrical generation in the former USSR were considerably higher than for Germany and UK, but that is to be expected as the production of electricity was also far higher in the USSR than in the selected West European countries.

It is also necessary to be aware, however, of the causes of atmospheric emissions when comparing Russian and West European data. For example, further data on sulphur dioxide emissions for 1988 (Boehmer-Christiansen and Skea, 1991, p. 35) reveal that power stations accounted for some 1.35M tonnes within a total of 2.23M tonnes for Germany (or some 60 per cent of total) and some 2.62M tonnes within a total of 3.66M tonnes (or some 71 per cent of total) for the UK. For nitrogen oxide emissions, power stations accounted for 0.73M tonnes within a German total of 2.96M tonnes (or some 25 per cent of total), and some 0.79M tonnes within a British total of 2.47M tonnes (or some 32 per cent of total) (Boehmer-Christiansen and Skea, 1991, p. 59). For the former USSR, however, electrical power generation accounted for some 43 per cent and 59 per cent of total industrial SOx and NOx emissions, as cited previously from the IMF (1991) report. These apparent variations in proportion between the former USSR on the one hand, and Britain and Germany on the other, were probably caused by the mix of fuels used and their consequent effects on SOx production; the levels of SOx emissions from the metallurgical industries (25 per cent in the case of Russia); and the high levels of NOx emissions from transport in Western countries, whereas personal road transport in the former USSR was not so highly developed. Further data are required to confirm these conclusions, however, as much of the Soviet data referred to industrial emissions, whereas those of Germany and UK referred to total national emissions.

In conclusion, therefore, it may become more important to consider the local effects of high densities of atmospheric pollution caused by the large scale combustion of particular types of fuel, rather than just national levels (or 'budgets') of pollutant emission. National data may be imprecise at source, and relate to a large number of disparate power stations and geographical areas. Furthermore, local concentrations of pollutants from power stations may also need to be considered alongside pollutants from other industrial facilities within a region, and the effects of adjacent industrial conurbations and associated transport infrastructures.

Table 2.13 Comparative Emissions of SOx and NOx (Million Tonnes)

	SOx Emissions (1980)	NOx Emissions (1987)
Assumed Total Emissions		
UK	5.3[1]	2.6[6]
Germany	3.6[1]	3.1[6]
Former USSR	25.3-28.5[2]	11.2[7]
Assumed Power Station Emissions		
UK	3.8[3]	0.8[8]
Germany	2.2[4]	0.8[9]
Former USSR	7.0-10.6[5]	2.7[10]

Notes:

1. See Table 2.1. Boehmer-Christiansen and Skea (1991), p 5 also quote a figure of 5.1M tonnes for UK SOx emissions in 1980 using EMEP data, and these same authors also provide (p. 240) an EC estimate of 4.7M tonnes for that same year. The corresponding figures quoted for West Germany during that same year are 3.6M tonnes (which is identical to that provided in Table 2.1 and provided above) and 3.2M tonnes. The higher quoted figures for both UK and Germany have been selected for comparative purposes in this table, as they are taken from the most recent publication of EMEP data.
2. See Table 2.5.
3. 5.3M tonnes (from Table 2.1) × 0.71 (as power stations accounted for 71 per cent of total UK sulphur dioxide emissions in 1988) (see Boehmer-Christiansen and Skea, 1991, p. 35).
4. 3.6M tonnes (see Table 2.1) × 0.6 (as power stations accounted for some 60 per cent of German sulphur dioxide emissions in 1988) (see Boehmer-Christiansen and Skea, 1991, p. 35).
5. See text.
6. See Table 2.6 (footnote).
7. See Table 2.9.
8. 2.6M tonnes (see Table 2.6) × 0.32 (as power stations accounted for some 32 per cent of total UK nitrogen oxide emissions in 1988). Boehmer-Christiansen and Skea (1991, p. 35) quote a 1986 figure from the Department of Environment of 2.5M tonnes of NOx emissions in total for the UK, including 0.8M tonnes from power stations. Boehmer-Christiansen and Skea (1991, p. 238) also quote a 1987 NOx emission figure for UK of 1.0M tonnes from 'existing plant' apparently based on estimates from 1980, but it is not clear whether that figure refers to power stations only, as power stations and industry combined. The 0.8M tonnes estimate as calculated above has therefore been chosen as the most reliable estimate for 1987.
9. 3.1M tonnes (see Table 2.6) × 0.25 (as power stations are reported to have accounted for some 25 per cent of total German nitrogen oxide levels in 1988), see Boehmer-Christiansen and Skea (1991, p. 35). Boehmer-Christiansen and Skea (1991, p. 35) quote a 1988 figure from the Bundesministerium fur Umwelt of 3.0M tonnes of NOx emissions in total for the Federal Republic of Germany, including 0.7M tonnes from power stations. Boehmer-Christiansen and Skea (1991, p. 238) also quote a NOx emission figure for the Federal Republic of Germany of 0.9M tonnes from 'existing plant' apparently based on estimates from 1980, but it is not clear whether that figure refers to power stations only or power stations and industry combined. The 0.8M tonnes estimate as calculated above has been taken as the figure for 1987.
10. See text.

POLICIES FOR REDUCTION OF ENVIRONMENTAL POLLUTION IN THE FORMER USSR

Pre-1991

The *glasnost'* policies of the Soviet government introduced in 1985 permitted a more open discussion of the environmental problems faced by the former USSR, and this process was reinforced by information gathered on the scale of radiation following the Chernobyl accident and from previous nuclear testing locations in Kazakhstan. In addition, a State Committee for Environmental Protection *(Goskompriroda)* was established in 1988, to submit proposals to the State Planning Committee *(Gosplan)* for environmental protection measures to be included as targets in five year plans for industrial ministries and their subordinate factories. Each republic was then to establish counterpart committees with subordinate offices in municipalities and local regions. Following the reform of the parliamentary system in 1989, a Committee on Ecology and the Rational Use of Natural Resources was created with responsibility to the USSR Supreme Soviet (IMF et al., 1991, p. 10).

Prior to the establishment of these bodies, decisions with regard to environmental pollution rested mainly with industrial ministries. Under the central planning system in operation at that time, and the dominance of incentives related to volume of output, the industrial ministries utilised the majority of their resources to the increasing of output rather than to meet any environmental concerns (Whitefield, 1993, pp. 158-62, 236-43). Furthermore, the censorship in operation in the USSR at that time stifled discussion of any concerns over pollution, and hindered the development of an environmental movement.

The implementation of measures proposed by *Goskompriroda* was hindered by bureaucratic inertia, difficulties in the recruitment of adequate staff, and continuous tensions with industrial ministries over powers and responsibilities, together with concerns that pollution control could lead to higher costs and consequent plant closure and unemployment. *Goskompriroda* also advocated the introduction of economic incentives based on fines and the use of the pricing mechanism, but the former was opposed by industrial ministries, whilst the latter was difficult to introduce because of the lack of a realistic pricing system from which to start (IMF et al., 1991, p. 12). Furthermore, although some 2 per cent of total state investment was allocated to environmental expenditures in 1990, which represented a 25 per cent increase from the 1988 budget, it was questionable whether the level of investment was sufficient for the scale of the problem. A major backlog existed from the lack of environmental concerns in the previous central planning system, and also as a failure to meet 1988 targets since only 55-65 per cent of air and water pollution targets were achieved (IMF et al., 1991, p. 14).

The majority of pollution control systems introduced were apparently 'end of pipe' control systems, and such designs are not usually as efficient or effective as process control throughout the various stages of production (IMF et al., 1991, p. 14). The USSR also became more engaged in international co-operation through frameworks of bilateral and multilateral agreements, including programmes with the United States on the reduction of environmental pollution (IMF et al., 1991, pp. 18-20).

One positive feature of environmental protection in the former USSR, however, was the drafting in 1989 of a state standard on emission limits from boilers (GOST 28269-89), containing data on NOx emission limits as shown in Table 2.14. The maximum emissions for the tabulated fossil fuels appear to compare favourably with European directives for new power stations of greater than 500MW capacity, but the Soviet data refer to the acceptance testing of new boilers rather than power station emissions in practice. Another set of data is also available, however, for recommended limiting values of boiler emissions from equipment supplied with gas cleaning equipment by the Ministry of Heavy Engineering of the USSR (*Mintyazhmash SSSR*) to the Ministry of Power of the USSR (*Minenergo SSSR*). These limits were subsequently approved by *Goskompriroda* in 1989 (see Table 2.15).

It can be seen from a comparison of the data provided for NOx emissions in Tables 2.14 and 2.15 that variations existed between the state standard and the *Goskompriroda* approved emission limits even for coals having low nitrogen contents. There also appears to be a difference in the assumptions made for NOx emissions in Tables 2.14 and 2.15, as permissible NOx levels increase as boiler capacity increases in Table 2.14, whereas NOx levels decrease or remain unchanged for most values in Table 2.15 as boiler capacity increases. These data suggest the necessity of fitting post-combustion equipment for large boilers burning heavy fuel oil and solid fuels, particularly those burning hard coal; although the permissible NOx limits may also be achievable by the use of low NOx burners, and thereby achieved in a more cost-effective way at the combustion stage. The widespread implementation of low NOx burners, however, would require reductions in the allowable NOx emission limits from the boilers themselves. GOST 28269-89 is currently being modified to include emissions from both boilers and power generation sets, and this document may subsequently promote the development of low NOx burners.

Post-1991

There is little evidence to suggest that environmental conditions have improved in the former constituent republics since the fragmentation of the USSR in 1991, except for reduced atmospheric emissions caused by falling industrial output and reduced demands for electricity. Each republic has had to establish its own system of government, although the Russian Federation was able to assimilate most of the former Soviet governmental structure because of its location in Moscow. Furthermore, the two largest industrial republics - Russia and Ukraine - have faced constitutional problems related to the relative powers of President and Parliament which have hindered the introduction and implementation of a wide range of legislation. Finally, the severe economic constraints created by budget and foreign trade deficits (although this latter factor may be less of a problem for Russia because of its large natural resource base), together with extremely high rates of inflation (Sakwa, 1993, pp. 201-49) have acted as severe limitations to investment in environmental control equipment, or the import of such equipment from the West. Responsibility for environmental protection and resource utilisation subsequently received ministerial status in the Russian Federation, but there is little evidence to suggest that more resources have been made available for the reduction of atmospheric emissions. However, Russia has now introduced more stringent limits for atmospheric emissions of SOx and NOx from new coal-fired power stations (200 mg/Nm^3) (Clarke, 1996, p. 56) which will clearly require attention to be paid to improve existing expertise in coal-fired combustion processes (see Chapters 3 and 4).

CONCLUSIONS

The research described in this chapter has been directed towards an assessment from published sources of the total levels of atmospheric pollution in the former USSR from oxides of sulphur and nitrogen, and the levels of these pollutants emitted by the electrical power generation sector which is regarded as the single largest industrial atmospheric polluter. The

Table 2.14 *Maximum Allowable NOx Emissions from Boilers (mg/Nm^3) (to GOST 28269-89)*
(Dry gas, $\propto$ = 1.40, t = 0^0C, 760mm. stpt)

Fuel	For boilers developed before 1/7/90		For boilers developed after 1/7/90	
	Steam capacity tons per hour			
	Less than 420	420 and higher	less than 420	420 and higher
Hard coal with solid slag removal nitrogen content (%/MJ per kg)				
less than or equal to 0.04%	470	550	470	550
more than 0.04%	570	650	570	570
Hard coal with liquid slag removal	640	700	640	700
Brown coal with solid slag removal nitrogen content (%/MJ per kg)				
less than or equal to 0.05%	350	450	320	370
more than 0.05%	450	570	350	450
Fuel oil	290	350	290	350; 250[1]
Natural gas	255	290	200	240; 125[2]

Notes:

1. For oil of mark 100 or better quality.
2. For boilers developed after 1/1/92 GOST 28269-89 also gives emission limits of 800mg/Nm^3 for boilers of capacity greater than 420 tons per hour (irrespective of date of development), for a fuel defined as 'ASh with liquid slag removal.' The author is uncertain as to whether this fuel refers to slack (*shlam*) or clinkered coal (*shlak*), possibly containing anthracite.

Table 2.15 Limiting Values of Pollutant Emissions from Boilers fitted with Gas Cleaning Equipment

Type of Fuel and Method of Combustion	Normative Concentration of Emissions mg/m^3			
	Steam Production Capacity in tons/hr or (Thermal Capacity MW)			
	> 420 (300)		< 420 (300)	
	SO$_2$	NOx	SO$_2$	NOx
Gas	-	125	-	250
Fuel Oil	400	185	600	290
Brown Coal				
Solid Slag Removal	400	225	600	340
Liquid Slag Removal	400	225	600	445
Boilers with Fluidised Beds	400	400	600	400
Boilers with Circulating Beds	400	200	600	200
Hard Coal				
Solid Slag Removal	400	240	600	470
Liquid Slag Removal	400	480	600	515
Boilers with Fluidised Beds	400	400	600	400
Boilers with Circulating Beds	400	200	600	200

Note: Concentrations of oxides of sulphur and nitrogen in the emitted gases are based on dry gases and reduced to $\propto$ = 1.4 (O$_2$ = 6%).
Source: Veshnyakov et al. (1990).

chapter has presented the data existing on levels of SOx and NOx pollution in the former USSR from government and international organisations, and the results from independent researchers using a variety of assumptions and methodology to arrive at their conclusions.

The range of estimates presented for total national emissions (between 16.2M and 28.5M tonnes of SOx in 1980, and between 4.5M and 11.2M tonnes of NOx in 1987) (see Tables 2.5 and 2.9) suggests that caution should be exercised in the interpretation of these data; but it is clear that even at the lower levels of estimates, Soviet emissions have been far in excess of West European countries (for example 5.3M tonnes of SOx and 2.6M tonnes of NOx for the UK, and 3.6M tonnes of SOx and 3.1M tonnes of NOx for West Germany) (see Tables 2.1 and 2.6). When the upper levels of Soviet estimates are considered, the disparities become even higher. From the viewpoint of SOx emissions, therefore, the former USSR was probably approximating to the level of the US (23.8M tonnes in 1980), but was far lower than its American counterpart in terms of NOx emissions (18.6M tonnes of NOx in 1987) even when the higher Soviet estimate (11.2M tonnes in 1987) is taken as a basis for comparison.

The disparities in atmospheric emissions are also apparent for the power generation sector, namely 7.0-10.6M tonnes of SOx for the former USSR compared with 3.8M tonnes for UK and 2.2M tonnes for West Germany in 1980; and 2.7M tonnes of NOx for the former USSR compared with 0.8M tonnes for both UK and West Germany (see Table 2.13). There is also evidence of high levels of SOx and NOx emissions from stationary sources in specific industrial conurbations.   Furthermore, although Soviet air emissions have been distributed over a wide area, there is growing evidence of consequent environmental damage to water resources and forests in a country which relies heavily on natural resource-based products (e.g. timber) for export purposes. In addition, transboundary pollutants from the former USSR have had a serious effect on the natural environments on several of their European neighbours, and the constituent republics of former USSR are likely to face similar international pressures to reduce SOx and NOx emissions from individual industrial facilities.

NOTES

1.   Weyman-Jones (1989), p. 32 comments that 'It is apparent that fossil fuel plant emissions are more related to the kWh energy supplied than to the maximum instantaneous peak demand deliverable by a given plant. Indeed, the ability to deal with emissions may be positively related to the size of the plant.'

2.   That is, for 1980 European Russia accounted for some 7.161M tonnes (see Table 2.3) of the 10.5M tonnes (see Table 2.2) emitted by the Russian Federation.
3.   Pryde (1991), p. 17, quoting a figure of 17.6M tonnes from the Report on the State of the Soviet Environment in 1988 *(Doklad: Sostoyanie prirodnoi sredi v SSR v 1988 godu)* published by *Goskompriroda* in 1989. The annual statistical handbook for the Soviet economy *(Narodnoe khozyaistvo SSSR v 1988g*, p. 249) reported a SOx emission level of 18M tonnes for that year.

REFERENCES

Bezuglaya, E. Yu. (1995), 'Air pollution in cities', in Feshbach, M. (ed.), *Environmental and Health Atlas of Russia,* Moscow: PAIMS, pp. 2-15, 2-16, 2-17.

Boehmer-Christiansen, S. and Skea, J. (1991), *Acid Politics: Environmental and Energy Policies in Britain and Germany*, London: Belhaven.

Clarke, L.B. (1996), *Coal Prospects in Russia: Perspectives* (IAPER/27) London: IEA Coal Research.

Derwent, R.G. (1985), *The Nitrogen Budget for the UK and North West Europe,* (ETSU Report R37, fig. 36, 1985.)

Dovland, H. (1987), 'Monitoring European Transboundary Air Pollution' , *Environment*, Vol. 29, No. 10.

Economic Commission for Europe (ECE) (1987), *National Strategies and Policies for Air Pollution Abatement*, ECE/EB.AIR/14, Geneva.

Economic Commission for Europe (ECE) (1994), Executive Body for the Convention on Long-Range Transboundary Air Pollution, *1994 Major Review on Strategies and Policies for Air Pollution Abatement: Tables and Figures*, (EB.AIR/R.87/Add.1), Geneva, 21 September 1994.

Goskomstat (1990), *Naronoe khozyaistvo SSSR v 1989 godu: staticheskii ezhegodnik*, Moscow: Finansy i statistika.

Gushcha, V.I. (1993),'Puti resheniya ekologicheskikh problem na ugol'nykh elektrostantsiyakh Rossii', Paper presented at the Symposium on New Coal Utilization Technologies, Helsinki, 10-13 May 1993, Economic Commission for Europe, Committee on Energy, Working Party on Coal, Geneva.

International Monetary Fund (IMF), the World Bank, the Organization for Economic Cooperation and Development (OECD), the European Bank for Reconstruction and Development (EBRD) (1991), A *Study of the Soviet Economy, Volume 3*, Paris: OECD.

Lyalik, G.N. and Reznikovskii, A.Sh. (eds) (1995), *Energetika i priroda,* Moscow: Energoatomizdat.

Mylona, S. (1993), *Trends of Sulphur Dioxide Emissions, Air Concentrations and Depositions of Sulphur in Europe since 1980*, EMEP/MSC-W report 2/93, Norwegian Meteorological Institute, Oslo.

Novozhilov, I.A. and Evdokimova, S.T. (1995), 'Ispol'zovanie prirodnogo gaza na elektrostantsiyakh i okhrana prirody', *Energetik,* No. 7, pp. 6, 7.

Pacyna, J.M., Larssen, S.R. and Semb, A. (1991), 'European survey for NOx emissions with emphasis on Eastern Europe', *Atmospheric Environment*, No. 25A (1991), pp. 425-39.

Pryde, P.R. (1991), *Environmental Management in the Former Soviet Union*, Cambridge: Cambridge University Press.

Ramus, C.A. (1991), *The Large Combustion Plant Directive: An Analysis of European Environmental Policy*, Oxford Institute of Energy Studies, Oxford (Working Paper EV7).

Ryaboshapko, A. (1990), 'Emission of pollutants in the USSR and their environmental effects', *Proceedings of Power Plant and Environment '90: the Greenhouse Effect and the Regional and Global Effects of Emissions*, Tampere, Finland, 31 October-2 November 1990.

Sakwa, R. (1993), *Russian Politics and Society,* , London: Routledge.

Tuovinen, J. P., Barrett, K. and Styve, H. (1994), *Transboundary Acidifying Pollution in Europe: Calculated Fields and Budgets 1985-93*, Det Norske Meterologiske Institut Technical Report No.129 (EMEP/MSC-W Report 1/94), Oslo, July 1994.

Tuovinen, J. P., Laurila, T., Lättilä, H., Ryaboshapko, A., Brukhanov, P. and Korolev, S. (1993), 'Impact of the sulphur dioxide sources in the Kola Peninsula on air quality in northernmost Europe', *Environment*, No. 27A, pp. 1379-95;

Veldt, C. (1991), 'Emissions of SOx, NOx, VOC and CO from East European countries', *Atmospheric Environment*, No. 25A, pp. 2683-700.

Veshnyakov, E.K., Varfolomeev, Yu. I. and Dmitrieva, R.L. (1990), 'Vybor sposobov ochistki dymovykh gazov energeticheskikh kotlov ot oksidov sery i azota', *Tyazheloe mashinostroenie*, No. 9, pp. 15-18.

Weyman-Jones, T.G. (1989), *Electricity Privatisation,* Aldershot: Avebury.

Whitefield, S. (1993), *Industrial Power and the Soviet State,* Oxford: Clarendon Press.

3. Fuels and Combustion

INTRODUCTION

As outlined in Chapter 2, the former USSR has been a major international source of atmospheric emissions of oxides of sulphur and nitrogen, and the electricity production sector has been the largest industrial contributor to atmospheric pollution in that region. As fossil-fuelled power stations accounted for some 70 per cent of electric power-generating capacity in the former USSR in 1991, and some 74 per cent of the electricity produced during that same year,[1] the levels of atmospheric pollution in the region have been influenced by the quantities and proportions of fuels burnt, their chemical properties, and the processes used for their combustion.

This chapter provides information on the chemical and combustion properties of fuels used in power generation facilities in the former USSR, paying particular attention to coal and heavy oil as the atmospheric emissions from these fuels are far higher than for natural gas. Reference is also made to the basic features of the combustion processes used to extract thermal energy from these fuels in order to provide kinetic energy for turbo-generation. Chapter 4 then extends this information by discussing the technologies available for reducing the emissions from conventional combustion technologies and the newer processes presently being introduced. The information in that chapter is then used for further discussion in Chapter 5, related to the fuel mix and power station capacities in the former USSR.

It is unlikely that the proportion of total power generated in thermal power stations in the region will change significantly in the foreseeable future, compared to the recent past. Although nuclear energy has been an important source of power in those European regions of the former USSR where fossil fuel delivery costs have been high,[2] it has not been a major source of electricity generation throughout the former Soviet Union as a whole, accounting for only some 10 per cent of Soviet power generation capacity in the early 1990s, and some 13 per cent of electricity produced.[3] Furthermore, concerns over safety aspects related to the design and operation of nuclear power stations in the former USSR, have led to a moratorium on the building of new units in many parts of the region (Kushnarev, 1994; IEA, 1994, pp. 125, 139). The future development of nuclear power in the former USSR will therefore probably be viewed primarily as a means of meeting capacity shortfalls in thermal and hydro-power facilities and consequently be significantly influenced by strategic decisions relating to the extraction, transport and combustion of fossil-fuels.

Hydro-electricity power generation technology is cleaner than either thermal or nuclear methods, and has accounted for some 19 per cent of electricity generation capacity, and some 14 per cent of electricity produced.[4] The further application of this technology is limited by the volumes and levels of water resources available and the hydro-environmental consequences of some proposals. Nevertheless, hydro-power presents a viable alternative for electricity power generation in many regions of Russia particularly in Siberia (Lagerev and Khanaeva, 1993) and especially in some of the Central Asian republics such as Kirghizia (Tuleberdiev, 1994). The development of hydro-power in the former USSR has therefore been considered alongside the development of fossil fuels within the research described in this book and is discussed in more detail later in Chapter 4.

Electricity generation from the combustion of industrial and domestic solid wastes, and the use of wind power, have not been discussed in this chapter as the utilisation of these techniques in the former USSR has been negligible.

FUEL AVAILABILITIES

Coal has been a major traditional fuel for electricity generation in the former USSR, and during the 1980s approximately 30 per cent of electricity from thermal power stations was generated using solid fuels, and there is a potential for this proportion to increase to 38 per cent.[5] Reserves of between 200 and 260G tons in the 'higher productive' and 'economically feasible' categories, have been located in the region, which could provide domestic energy for another 290-350 years at the output levels of some 700 million tons achieved in the 1980s (Hewett 1984, p. 28).

Some other estimates are available which suggest that coal reserves are even more plentiful, exceeding some 6000G tons (Vdovchenko, 1993), comprising 2,000G tons of brown coal to depths of 600m and hard coal reserves of more than 4,000G tons to depths of 1800m (Dienes and Shabad, 1979, pp. 105-7), although the problems of extraction of some of these reserves, combined with their remote location and associated transport costs to major customers, raise questions of technical and economic viability. Furthermore, as a result of high transport costs over the long distances from coalfields to ports, the export sales of coal to the OECD countries (some $0.4bn in 1992) have not been high when compared with oil and gas (some $14bn of mineral fuels, lubricants and related materials and some $4bn of natural gas in that same year). Increased use of coal on the domestic power generation market consequently creates a possibility of more oil and gas being available for export, and provides a critical outlet for the coal mining industry absorbing some 40 per cent of that industry's output in the past (Hewett 1984, p. 107).

Oil, the traditional hydrocarbon fuel of the former USSR, has been cheaper to transport than coal (Elliot, 1974, pp. 217-19) and has been a far higher earner of hard currency income, accounting for more than 50 per cent of Soviet earnings from Western countries (Hewett, 1984, pp. 153-9). Available oil reserves compared with consumption are far more modest than those of coal, however, as present known reserves may only last until the beginning of the twenty first century unless there is further investment in exploration and extraction technology (Hewett, 1984, p. 29). As exports have accounted for some 20-30 per cent of oil production (Stern, 1993, p. 99), it remains likely that oil will only be viewed as a secondary fuel for electricity production as its share of fossil fuel production for Russian electricity and heat generation has fallen from 40 per cent in the 1980s to 13 per cent in 1992 (IEA, 1994, p. 199). Oil extraction difficulties have been gradually increasing, contributing to a steady decline in production of that fuel from 624 million tons in 1988 to 449 million tons in 1992, although some decline in production may also have been caused by falling requirements as industrial output has decreased.

Soviet resources of natural gas were estimated to account for 40 per cent of the world's reserves of this fuel in the 1980s and were considered capable of being exploited until the middle of the twenty first century (Hewett, 1984, p. 28). A complex system of pipelines has been established for gas transportation, and natural gas now accounts for some 50 per cent of the fossil fuels used in power generation in Russia, which is almost double the proportion of the early 1980s (Kudravyi, 1994; IEA, 1994, p. 34).

SOLID AND LIQUID FUELS

Combustion Properties

In conventional combustion processes, either pulverized coal or oil is burnt in a boiler to heat water to drive a steam turbine. SOx and NOx emissions are transferred to the atmosphere via the boiler stack (see Figure 3.1), and the quantity of these emissions can vary with the fuel characteristics and the type of in-combustion or post-combustion control. This chapter will consequently present information on the characteristics of solid and liquid fuels available in the former USSR, and the following chapter will discuss modifications to conventional combustion processes, and advanced combustion technologies which can reduce SOx and NOx emissions.

A mix of coals, including anthracite, hard coal (bituminous coal) and brown coal (lignite) is used at Russian power stations, with brown coal and hard coal accounting respectively for some 50 per cent and 47 per cent of the total coal burnt (see Table 3.1). Anthracite only accounts for some 3 per cent of fuel burnt, although this fuel has a higher average calorific value (5070 kcal/kg) than either hard coal (4480 kcal/kg) or brown coal (3050 kcal/kg). Better quality fuels have traditionally been delivered to the metal processing industries, partly because the high humidity, ash content and poor coking quality of brown coals renders them unsuitable for that area of application. Power stations have consequently had to use coal of lower quality in terms of their calorific value.

The aggregated data provided in Table 3.1, however, hides the variations in calorific values for both hard coals and brown coals as shown in Table 3.2, presumably defined from standardised test conditions. In the case of hard coal for example, both Kuznets and Nerungrinsk coals have a higher calorific value than Donets coal mined in Ukraine and south-west Russia, whilst the calorific values of Ekibastuz (Kazakhstan) and Cheremkhovsk coals are only some 3850 kcal/kg and 3730 kcal/kg respectively. For brown coals, there are a few varieties which approach the calorific values of some hard coals. Azeisk coal for example has a calorific value of 3880 kcal/kg and Kansk-Achinsk coal has a calorific value of 3,600 kcals/kg, but the calorific values of other varieties are as low as 1720 kcal/kg (Bikinsk coal) and 1990 kcal/kg (Moscow coal). The variations in calorific value evident in different types of Russian coal are also apparent for sulphur content, which varies from 0.4 per cent for Kansk Achinsk brown coal

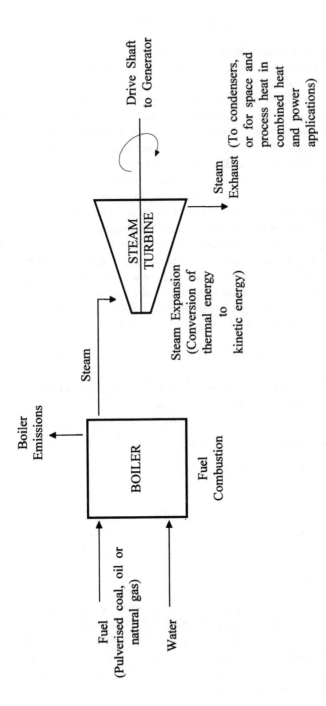

Figure 3.1 Conventional Fuel Combustion

to 3.2 per cent for Pechorsk hard coal (mined in the Arctic Circle), which has the highest sulphur content of all of the coal varieties listed in Table 3.2.

More apparent, however, are the different characteristics of the varieties of coal which consequently affect their combustion capabilities; and it is difficult to distinguish 'high performance' from 'low performance' coals, because of the range of fuel characteristics which can define performance and the interactions between them. Donets anthracite, for example, has a high calorific value, low humidity, and moderate ash content, but a high sulphur content of 1.9 per cent. In spite of this sulphur content, however, it is likely that some Donets coal will continue to be used for some power generation in the former USSR, in view of its high calorific value and close proximity to major industrial regions, although anthracite has not been used extensively in power stations and this source of coal is now becoming more difficult to extract. Kuznets hard coal, on the other hand, has approximately the same calorific value as Donets anthracite but a far lower sulphur content at 0.42 per cent, and can therefore be viewed as an attractive choice of fuel for coal-fired power stations; although its good coking properties also make it an attractive fuel for the metal processing industries.

Table 3.1 Quality and Quantity of Coals Fired at Power Plants (1991)

	Quantity million tonnes	Humidity $W_r\%$	Ash content A_d %	Heat of combustion Q_r	
				kcal/kg	MJ/kg
Coal, total	173.7	20.2	29.6	3800	15.9
including					
Anthracite	6.0	7.7	28.9	5070	21.2
Hard coal	81.6	9.0	33.4	4480	18.7
Brown coal	86.1	31.4	25.4	3050	12.8

Source: Vdovchenko (1993)

Variations are also apparent amongst the varieties of brown coal, with a high sulphur content, high humidity, high ash content and low calorific value for Moscow coal, compared with a lower sulphur content, higher humidity, lower ash content and higher calorific value for Kansk-Achinsk coal. All of the brown coals presented in Table 3.2, however, have a high volatility compared with the hard coals.

It is apparent, therefore, that the techniques selected for reduction of pollution emissions and the level of success of these techniques will probably vary significantly according to the variety of coal being burnt. For Moscow coal, which may still be used in some of the capital's coal-fired power stations because of the advantages of proximity, and also for Pechorsk coal as this fuel has been used near to the highly urbanised area of St. Petersburg (Elliot, 1974, pp. 108-9), it is apparent that attention is required to reduce sulphur emissions and volatility with the added problems of high ash and humidity in the case of Moscow coal.

At present, however, it appears that the majority of coal deposits in the former USSR are accounted for by the Kuznets and Kansk-Achinsk coalfields in Siberia and the Ekibastuz coalfields in Kazakhstan. According to the data shown in Table 3.2, Kuznets coal in particular has a high calorific value, combined with a comparatively low sulphur content and humidity and a moderate ash content. Combustion problems should not therefore be particularly high which may also increase its potential for sale in Western markets. Ekibastuz coal, however, although drier than the Kuznets variety, will require more attention to be paid to higher sulphur and ash contents. Kansk-Achinsk coal has low sulphur and ash contents, and also a comparatively high calorific value for brown coals, although this parameter is probably too low to justify transport over long distances. The humidity of Kansk-Achinsk coal, however, is extremely high, and it is prone to spontaneous combustion if exposed to air (Dienes and Shabad, 1979, p. 118).

Amongst these three varieties of coal, therefore, it appears that the sulphur content is not particularly high when compared with British varieties which have varied between 0.5 per cent and 2 per cent with an average of 1.6 per cent, although German hard coal has a slightly lower sulphur content than its British counterparts (Boehmer-Christiansen and Skea, 1991, p. 43). The Ekibastuz and Kansk-Achinsk coals have a lower calorific value than the Kuznets coals, however, and should consequently emit higher levels of sulphur dioxide per unit of heat produced; although the data provided in Table 3.3 suggest that the specific emissions of sulphur per unit of heat generated are lower for Kansk-Achinsk brown coal than for Kuznets hard coal, even though Table 3.2 shows that the sulphur contents are approximately the same, the calorific value of the Kuznets coal is higher, and the specific sulphur compositions per unit of heat generated were quoted as identical. It is therefore likely that the differences between SOx emissions and sulphur contents of Kuznets and Kansk-Achinsk coals as listed in Tables 3.3 and 3.2, are caused by differences in volatility, ash

Table 3.2 Properties of Coal Burned in Russian Power Stations

	Quality characteristics of coal				Heat of combustion		% kg/MJ	
	W_r [2] %	A_d [3] %	S_d [4] %	V_{daf} [5] %	kcal/kg	MJ/kg	A_p [6]	S_p [7]
(i) Anthracite								
Donets[1] (less than 128G tons)	7.7	28.9	1.9	4.0	5070	21.2	1.26	0.08
(ii) Other Hard (Bituminous) Coals								
Kuznets[1] (725G tons)	11.7	23.4	0.42	-	5160	21.6	0.96	0.02
Ekibastuz	5.9	43.4	0.6	25.0	3850	16.1	2.54	0.04
Pechorsk[1] (214G tons)	10.9	33.7	3.2	-	4170	17.4	1.72	0.16
Cheremkhovsk	17.8	32.3	1.2	6.0	3730	15.6	1.7	0.06
Nerungrinsk	8.8	19.3	0.2	20.0	5690	23.8	0.74	0.01
(iii) Brown coals (Lignite)								
Near Moscow[1] (20G tons)	30.2	48.4	3.0	50.0	1990	8.32	4.06	0.25
Kansk-Achinsk[1] (600G tons)	34.0	10.4	0.4	48.0	3600	15.05	0.46	0.02
Chelyabinsk	15.2	45.3	1.0	45.0	2950	12.34	3.11	0.07
Azeisk	24.0	21.7	0.6	48.0	3880	16.22	1.02	0.03
Bikinsk	38.3	45.9	0.4	53.0	1720	7.19	3.93	0.03
Kharanorsk	38.4	22.0	0.5	44.0	2720	11.37	1.19	0.03

Notes:
1. Data on available reserves
2. Humidity content
3. Ash content
4. Sulphur content
5. Reaction activity, or volatiles yield
6. Specific ash content
7. Specific sulphur content

Source: Vdovchenko (1993)

contents and combustion properties, but also differences in sulphur contents of the two selected sets of samples, as Table 3.4 shows three of the four varieties of Kansk-Achinsk coals to have a sulphur content of less than the 0.4 per cent figure quoted in Table 3.2. When units of power generated are selected as the denominator, however, the sulphur dioxide emissions of Kuznets coals are lower than those Kansk Achinsk coals, possibly because their combustion properties cause less boiler downtime, although the ash contents of Kuznets coals are listed as higher than the Kansk-Achinsk coals in Table 3.2 and this factor can affect boiler efficiency.

An important factor to note from Table 3.2, though, is the high value of sulphur content and low calorific value for brown coal from the Moscow region, and Table 3.2 also shows high sulphur emissions from Donets coals. Table 3.3 provides data on NOx emissions for a selected range of coal varieties, and these data also suggest that Kansk Achinsk brown coals have comparatively low levels of NOx emissions, although these emissions appear to be moderate to comparatively high for Kuznets hard coal. Furthermore, the data presented in Table 3.2 show the high levels of SOx and NOx emissions from fuel oil and the particularly high levels of SOx emissions from shale.

The data shown in Tables 3.2 and 3.3, however, are limited in terms of the potential for comparison with Western standards, as the emission data is not provided in mg/Nm^3 and very little account is taken of the fact that different combustion properties of fuels having similar sulphur contents may give rise to different levels of atmospheric emissions. The only figures obtained to date by the present author for both SOx and NOx emissions from coal-fired power stations in the former Soviet region, which can be used for comparison with Western practice, are those provided by Prutkovskii et al. (1993). This source refers to SOx emissions varying from 1900-2400 mg/Nm^3 from coal-fired power stations in the former USSR, and also provides data on NOx emissions varying from 450 to 2,200mg/m^3 although a range of 300 to 1,200mg/m^3 for NOx emissions is cited elsewhere (Fillipov, 1990).

The data provided in Tables 3.2 and 3.3 have been useful, however, to compare with more specific data on the thermal characteristics, and sulphur content and SOx emissions of various varieties of coal from particular coalfields in the former USSR as presented in Table 3.4. This table also presents data for SOx emissions in mg/m^3 for a range of solid and liquid fuels available in the former USSR, although it is not entirely clear whether all of these fuels are used in power generation in that region.

Table 3.3 SOx and NOx Emissions Data for Different Fuels

Parameter	Unit	Fuel					
		Shales	Coals				Fuel oil
			1	2	3	4	
Specific emissions:							
SO_2 per unit of heat	g/MJ	2.17	0.32	1.02	1.41	0.6	1.72
	kg/ton[5] of fuel equivalent	63.55	9.51	29.97	41.25	15.00	50.43
per unit of electricity generated							
SO_2	g/KWh	35.39	3.43	11.08	14.01	3.2	16.12
NOx	g/KWh	2.00	1.95	3.01	2.04	3.1	3.4

Notes: 1. Kansk-Achinsk coals.
　　　　2. Ekibastuz coals.
　　　　3. Donets coals
　　　　4. Kuznets coals.
　　　　5. Tonnes of fuel equivalent to 7000 k.cal/kg or 29.33 MJ/kg heat capacity.

Source: Gushcha (1993)

The data shown in Table 3.4 confirms the moderate levels of sulphur content for most types of Kuzbass and Kansk-Achinsk coals, and the related moderate levels of SOx emissions for those coals which are common types of solid fuels used in Russian power stations. Some varieties of Neryungren coals may apparently be burnt without the use of additional SOx control techniques to meet the former USSR's State Committee of the Environment's (*Goskompriroda*) requirement of 600mg/Nm^3 for small boilers' SOx emissions. There are only five varieties of coal, however, that can approach the European Union's Large Combustion Plant Directive (LCPD) (see Chapter 1) and *Goskompriroda* requirement of 400mg/Nm^3 for large boilers without some form of coal cleaning combined with flue gas desulphurisation, namely Neryungren coal, one variety of Kuzbass coal and three varieties of Kansk-Achinsk coal, where advanced coal cleaning techniques alone may be sufficient if shown to be technically and economically feasible. In addition only the low sulphur version of the available liquid fuels burns with a level of SOx emissions acceptable to both the LCPD and *Goskompriroda* requirements for large power stations, and none of the solid or liquid fuels can meet the latest Russian legislation for new power station capacity without some form of flue gas desulphurisation.

It is also apparent that the levels of SOx emissions from some of the solid fuels listed in Table 3.4 are extremely high, particularly for Ukrainian coals (both Donbass and L'vov-Volyask) and Leningrad and Estonian shales. The reduction of sulphur dioxide emissions from shale-fired power stations in Estonia and North-West Russia would therefore appear to be an urgent priority, particularly as most of these power stations are close to the Finnish border and consequently attract international concern.

Coal Cleaning and Oil Refining

The sulphur content of coal can be reduced by coal cleaning, as up to half of the sulphur content of coal is in pyritic form which can be removed using a physical process which exploits the difference in densities between combustible material and ash. Standard coal cleaning can reduce SO_2 emissions by 10-30 per cent (IEA, 1988, p. 69), but removal of the remaining organic sulphur from coal would require chemical or biological processing at a much higher cost (Boehmer-Christian and Skea, 1991, p. 44), although advanced chemical coal cleaning techniques currently being developed may be able to achieve 90 per cent reductions (IEA, 1988, p. 69). The possibility of using coal cleaning systems within Eastern Europe is demonstrated by a new cleaning unit under construction in Poland with

Table 3.4 *Thermal Characteristics, Sulphur Contents and SOx Emissions of Various Soviet Coals*

Area or Republic	Basin	Fuel Code	Class of Dressing	Minimum Combustion Value MJ/kg	Sulphur Content %	SO_2 dry gas (mg/m^3) $\alpha = 1.4$
Ukraine	Donets (Donbass)	D	Screenings	16.95	3.5	11,170
		G	Screenings	18.88	3.5	10,200
		G	Industrial coal	15.95	2.9	9,930
		G	Slurry	14.44	2.5	9,360
		Zh,K,OS	Industrial coal	17.00	2.8	8,980
		Zg,K,OS	Slurry	16.70	2.5	8,160
		T	R	23.40	2.4	5,590
		A	Slack, GSh	19.97	1.6	4,340
		A	Slack	15.09	1.4	4,860
Kemerovsk	Kuznets (Kuzbass)	D	R,SSh	22.86	0.4	970
		G	R,SSH	25.25	0.5	1,110
		2SS	R,Screenings	24.70	0.4	890
		1SS	R,Screenings	23.57	0.3	700
		T	R,Screenings	25.12	0.5	1,090
		G	Industrial coal	20.01	0.5	1,360
		Zh	Industrial coal	18.76	0.7	2,020
		K,Zh	Industrial coal	19.85	0.3	810

	Open Cast	G	ROK I	22.94	0.3	720
		G	ROK II	18.63	0.3	880
		1SS	ROK I	23.99	0.4	920
		1SS	ROK II	18.21	0.3	890
		2SS	ROK I	23.94	0.3	680
Kazakhstan	Karaganda	K	Industrial coal	16.24	0.8	2,600
		K	Slurry	18.42	0.9	2,650
	Ekibastuz					
	2nd group	SS	R	14.44	0.8	3,090
	1st group	SS	R	17.38	0.7	2,230
	Turgaiskoe	B2	-	13.15	1.4	5,570
L'vov-Volyask	Volyask	G	R,Screenings	20.85	3.0	7,870
	Mezh-reshensk	G2h	R,Screenings	19.38	2.8	7,920
Krasnoyarsk	Kansk-Achinsk:					
	Irsha-Borodinsk	B2	R	15.49	0.2	690
	Nazarov	B2	R	13.02	0.4	1,590
		B2	R	15.66	0.2	680
	Berezov	B2	Low Ash	16.20	0.2	650

Table 3.4 continued

Area or Republic	Basin	Fuel Code	Class of Dressing	Minimum Combustion Value MJ/kg	Sulphur Content %	SO_2 dry gas (mg/m^3) $\propto = 1.4$
Irkustk	Azeisk	B3	R	16.91	0.4	1,300
Buryat	Gusino-ozersk	B3	R	16.16	0.7	2,330
Chitinsk	Kharanorsk	B1	R	11.97	0.4	1,720
Khabarovsk	Raichikhin	B2	K,O,MSSh,R	12.73	0.3	1,210
Yakut	Neryungren	SS	R	22.48	0.2	490
Estonia	Estonian Shale	Inflammable Shale	Mined, open cast	9.00	1.4	8,620
Leningrad	Leningrad Shale	Inflammable Shale	Mined, open cast	7.66	1.3	9,260
		Fuel Oil	Low sulphur	40.27	0.3	420
			Medium sulphur	39.73	1.4	2,000
			High sulphur	38.76	2.8	4,080

Source: Veshnyakov et al. (1990)

technical assistance from ECN (Netherlands), which aims to reduce pyritic sulphur content by 50 per cent. The average sulphur content of the coal will be some 1.5 per cent, from mined coal containing some 2.6 per cent of sulphur by weight in its untreated state (Oudhuis et al., 1993). The effects of the remaining levels of sulphur can be removed by attention to post-combustion processes, whilst reduction of NOx emissions can be achieved at the combustion and post-combustion stages.

Research carried out in the United States has demonstrated a correlation between conventional coal cleaning and improved performance for pulverised coal-fired boilers. Benefits included reduced boiler erosion, mill wear and slagging, and increased mill capacity; but specific attention had to be paid to the particular requirements of low-ranking coals in view of the wide variability in their washability (Sondreal, 1992). US technical and economic assessments of advanced fine-coal-cleaning methods for sulphur control concluded that it is only marginally feasible to meet a 1.2lb/MMBtu SO_2 standard by coal cleaning alone and only for selected coals containing a high per centage of pyrite versus total sulphur. The 1.2lb/MMBtu SO_2 standard was the 1990 US federal commission standard for SO_2 emissions, which approximated to $1600mg/Nm^3$, or some four times higher than the European Union's Large Combustion Plant Directive for power stations larger than 500MW. Although the combustion and emission properties of coals can vary according to the proportions of pyritic and total sulphur content, the sulphur content of a coal producing $400mg/Nm^3$ of emissions would be of the order of 0.15-0.2 per cent sulphur (Sondreal, 1992), assuming favourable combustion properties.

Referring again to the technical data for Russian coals having sulphur contents lower than 0.2 per cent provided in Tables 3.2 and 3.4, it is again apparent that even if coal cleaning is technically and economically feasible to meet SOx emissions of $400mg/Nm^3$, it may only be applicable for Nerungren coal, and some low sulphur varieties of Kuzbass and Kansk-Achinsk coals. For other types of coal, however (such as Ekibastuz, Azeisk, Gusinoozersk, Kharanorsk, Raichikhin and some of the higher sulphur Kuzbass coals), it will be necessary to combine coal cleaning with some form of post-combustion treatment (such as flue gas desulphurisation) or a newer advanced coal combustion technology such as fluidised bed or gasification (see Chapter 4). In the cases of Ukrainian coals from the Donbass, L'vov and Volyansk regions, however, and inflammable shales from the Leningrad and Estonian regions in the North West, it would appear that their SO_2 emissions may be beyond the levels that can be reduced to the

LCP Directive, even when using post-combustion treatment or advanced coal technologies.

It is also apparent that oil refining can substantially reduce the sulphur content of fuel oils (see Table 3.4), which may be a more economic option than desulphurisation of boiler emissions.

NATURAL GAS

From the data presented in Tables 3.2-3.4, it can be seen that the levels of SOx and NOx emissions are comparatively high for solid and liquid fuels used for power generation in the former USSR. The most desirable fuel from the environmental viewpoint is therefore natural gas, because it is almost sulphur free, and burns with NOx and CO_2 emissions some 30 per cent lower than oil and less than half those for coal, as it contains nitrogen in molecular form only and not as compounds.[6] Another factor favouring the use of natural gas as a fuel in power generation is convenience of transport, and consistent calorific value compared with Soviet coal's 23 per cent reduction in this parameter between 1965 and 1982 (Hewett, 1984, p. 107) with no evidence of any improvement since that date.

Most gas fired power stations in the former USSR have operate gas-fired boilers to heat water for steam turbo-generation (See Figure 3.1). Gas turbines have been used for response to peak load conditions over a number of years, and are now being more widely used in base load generation. Exhaust gases from gas turbine cycles can also be used for steam generation purposes, and such 'combined cycles' can yield high levels of efficiency (see Chapter 4).

CONCLUSION

This chapter has presented information on the fuels available for power generation in the former USSR and the diversity of their combustion properties. Particular attention has been paid to the high levels of SOx emissions of some solid fuels and heavy fuel oils, and the higher rates of fuel consumption which may be required to compensate for the low calorific value of certain types of brown coal.

Natural gas has been presented as the favoured fuel in the former USSR in view of its high efficiency of combustion, comparatively low transport costs, negligible SOx emissions, and low values of NOx emissions; although coal still remains an economic option in some locations and further investigations should consequently be carried out on the feasibility of fuel cleaning at selected coal fields (for example, Western Siberia). Furthermore, as natural gas and oil have a higher export potential than coal in view of their lower transport costs, further opportunities could be thereby created for the sale of coal to domestic power stations. Much of the data presented in Chapter 4 therefore focuses on the comparative efficiency and cleanliness parameters of both gas and coal combustion processes.

NOTES

1. In 1991, capacity in Soviet thermal power stations was some 243,000MW within a total power generating capacity of 344,000MW, and electricity output was some 1238TWh within a total of 1684TWh (see D'yakov et al., 1994).
2. Relative costs for fuels are provided by Lagerev and Khanaeva (1993). Stern (1995, p. 34) notes that 38 per cent of electricity generation capacity is nuclear-powered in the Russian North West, and 20 per cent in the Central region.
3. In 1991, capacity in Soviet nuclear power stations was some 36,000MW and electricity output was some 212TWh. (D'yakov et al.,1994).
4. In 1991, capacity in Soviet hydro-electric power stations was some 65,000MW, and electricity output from these units was some 234TWh. (D'yakov et al.,1994).
5. Hewett (1984, p. 107) provides a total primary energy (TPE) requirement of 24,352 thousand barrels per day of oil equivalent (tbdoe) for 1980, of which 31 per cent (or 7,549 tbdoe) was provided by solid fuels. 7147 tbdoe were used in the generation of electrical power, of which 43 per cent (or 3073 tbdoe) was provided by solid fuels. Consequently electricity generation accounted for some 41 per cent (3073/7549) of the solid fuels TPE. Although the category 'solid fuels' included a range of materials (such as peat, firewood and oil shale - an important local fuel in the Russian North West), coal was the dominant fuel. Furthermore, thermal power stations accounted for some 80 per cent of electricity production in 1980 of which 43 per cent, were heated by solid fuels, thereby providing 34 per cent (0.8 × 0.43) of fuels for electricity production. In 1992, coal accounted for some 27 per cent of the Russian fossil fuel input for heat and electricity generation, and the Russian authorities hoped to increase this proportion to 38 per cent in the future to release more oil for export. (IEA, 1994, p. 199).
6. The data for SOx and NOx emissions are cited from Boehmer-Christiansen and Skea, (1988, p. 44) and IEA (1988, p. 29). The data for CO_2 emissions are cited from Suvorina (1995) (that is, 2300kg of CO_2 per 1120m^3 of natural gas, compared with 3100kg per ton of coal, and 4800kg per ton of oil). Suvorina quotes Doelman (1991) as her source for the cited emissions data.

REFERENCES

Boehmer-Christiansen, S. and Skea, J. (1991), *Acid Politics*, London: Belhaven.

Dienes, L. and Shabad, T. (1979), *The Soviet Energy System : Resource Use and Policies*, Washington, DC: V.H. Winston and Sons.

Doelman, J. (1991), 'Natural Gas Bridge to a Clean Energy Future', *Proceedings of the World Energy Conference*, Geneva, 4-7 November 1991, pp. 181-96 .

D'yakov A.F., Dzhangirov, V.A. and Barinov, B.A. (1994), 'Problemy ko-ordinatsii razvitiya i funktsionirovaniya ob''edinenykh energosystem Sodurzhestva Nezavisimykh Gosudarstv', *Energetik*, No. 2, pp. 2-6.

Elliot, I.F. (1974), *The Soviet Energy Balance: Natural Gas and other Fossil Fuels,* New York: Praeger.

Fillipov, G.A. (1990), 'Ekologicheskie aspekty v energetike i mashinostroenie', *Tyazheloe mashinostroenie*, No. 9, pp. 2-6.

Gushcha, V.I. (1993), 'Puti resheniya ekologicheskikh problem na ugol'nykh elektrostantsiyakh Rossii', Paper presented at the Symposium on New Coal Utilization Technologies, Helsinki, 10-13 May 1993, Economic Commission for Europe, Committee on Energy, Working Party on Coal, Geneva.

Hewett, E. (1984), *Energy, Economics and Foreign Policy in the Soviet Union*, Washington DC: Brookings Institution.

International Energy Agency (IEA) (1988), *Emission Controls in Electricity Generation and Industry*, Paris: OECD.

International Energy Agency (IEA) (1994), *Electricity in European Economies in Transition*, Paris: OECD.

Kudravyi, V.V. (1994), 'O perspectivakh nauchno-tekhnicheskogo progressa v elektroenergetike', *Energetik*, No. 3, pp. 2, 3.

Kushnarev, F.A. (1994), 'Aspekty reorganizatsii sistemy Rostovenergo', *Elektricheskie stantsii*, No. 6, pp. 16-19.

Lagerev, A.V. and Khanaeva, V.N., (1993) 'Stsenarii razvitiya Energeticheskogo Kompleksa Rossii', *Energetik*, No. 11, pp. 5-8.

Oudhuis, A.B.J., Yansen, D., Iwanski, Z. and Golec, T.W. (1993), 'Repowering of Polish coal-fired plants with IGCC', Paper presented at the Twelfth EPRI Conference on Gasification Power Plants, 27-29 October 1993, Hyatt Regency, San Francisco, USA.

Prutkovskii, E.N., Safonov, L.P., Varvarskii, V.S. and Borovskii, V.M. (1993), 'Povyshenie ekologicheskoi effektivnosti TES pri poetapnom sovershenstvovanii PGU s gazifikatsei uglya', *Teploenergetika*, No. 9, pp. 51-6.

Sondreal, E.A. (1992), 'Clean Utilization of Low-Rank Coals for Low-Cost Power Generation', *Proceedings of the Energy and Environment: Transitions in Eastern Europe Conference*, Prague, 20-23 April 1992, pp. 53-87.

Stern, J.P. (1993), *Oil and Gas in the Former Soviet Union,* London: Royal Institute of International Affairs (RIIA).

Stern, J.P. (1995), *The Russian Natural Gas 'Bubble': Consequences for European Gas Markets*, London: The Royal Institute of International Affairs, Energy and Environmental Programme.

Suvorina, L.I. (1995) 'Mirovaya energetika i okhrana okruzhayushchei sredy', *Energeticheskoe stroitelstvo*, No. 6, pp. 21-7

Tuleberdiev, Zh.T. (1994), 'Sostoyanie i perspektivy elektroenergetiki v Kirgyzkoi Respublike', *Energetik*, 1994, No. 5, pp. 3, 4.

Vdovchenko, V.S. (1993), 'Specific Features and Ecological Characteristics of Russian Coals Supplied to Power Stations', UNECE Committee on Energy, Symposium on New Coal Utilization Technologies, Helsinki, 10-13 May 1993.

Veshnyakov, E.K., Varfomoleev, Yu. I. and Dmitreva, R.L. (1990), 'Vybor sposobov ochistki dymovykh gazov energeticheskikh kotlov ot oksidov sery i azota', *Tyazheloe mashinostroenie,* no. 9, pp. 15-18.

4. Combustion Processes

INTRODUCTION

This chapter discusses the technological options available for the reduction of atmospheric pollution during the power generation process, within the framework of fossil fuel combustion systems, generation efficiency, and energy conservation.

Chapter 5 below provides evidence to demonstrate that although oil remains a fairly important fuel for power generation in North West Russia, the proportion of oil used as a fuel in domestic power generation in the former USSR has gradually decreased as a consequence of the potential for export sales of this commodity to the West, combined with continuing difficulties in oil extraction in many regions. Oil is therefore now viewed in many regions of the former USSR as a reserve fuel in case of shortages of gas, as an enrichment agent, or used for boiler ignition. It is therefore assumed in the remainder of this chapter that apart from North West Russia, investment in new oil-fired power stations is not likely to have a high priority in the former USSR, when compared with investments in gas-fired and solid-fuel fired power stations. The majority of attention in this chapter has therefore been devoted to the reduction of SOx and NOx from gas and solid fuel combustion, although some of these techniques can also be used in oil burning.

The remainder of this chapter is therefore divided into five separate sections, the first three of which describe the technologies available for the reduction of SOx and NOx from coal and gas respectively. Coal combustion technologies are considered first, as coal was originally used as the conventional fuel for electricity generation. Descriptions are provided of the in-combustion and post-combustion processes which have now become well established in industrial practices and can be viewed as 'conventional' technologies. That section is then followed by a description of gas combustion technologies, paying particular attention to the use of gas turbines and combined cycles which provide scope for improved efficiency from the use of this fuel, which is inherently cleaner than solid and liquid fuels.

A section on the newer advanced coal technologies is placed after the description of gas combustion as some newer coal combustion technologies can also be used to generate gas for use in turbine combustion, as well as reduced atmospheric emissions from the burning of solid fuel. The description of advanced coal technologies is then followed by a consideration of the options available for energy conservation in both power generation and power utilisation, and the chapter is concluded with a discussion of appropriate technologies for transfer to the former USSR.

Before commencing this description of the various combustion technologies, however, it is important to note that SOx reduction can be substantially reduced by appropriate treatment of fuels before combustion, as highlighted in Chapter 3. These alternatives, however, require investment in the fuel extraction and processing industries (for example, coal mining and oil refining) rather than power generation and power engineering with which we are chiefly concerned in this present chapter.

The following three sections of the chapter give a brief description of each of the relevant technologies for the reduction of SOx and NOx, and then discuss the potential for the use of those technologies in the former USSR. The description of the technologies has been compiled from a range of Western published sources describing both scientific concepts and practical applications. The potential for the use of these technologies in the former USSR has been established from a survey of technical publications in the Russian language relating to power generation and power engineering, and discussion with Western and Russian technical specialists with expertise in these fields.

CONVENTIONAL COAL COMBUSTION

Introduction

As outlined in Chapter 3 emissions from coal combustion are transferred to the atmosphere via the boiler stack (see Figure 3.1). The reduction of SOx and NOx emissions can therefore be achieved by treatment of these emissions before they are released to the atmosphere ('post-combustion systems'), or by closer specification and control of the combustion process ('combustion modification'). This chapter therefore describes the major features of these systems, drawing attention to their present levels of practicability.

Combustion Modification

Low NOx burners

Low NOx burners modify the combustion process in two ways, namely: a reduction of the flame temperature to below 1500°C, in order to minimise the atmospheric nitrogen reactions which are strongly temperature dependent; and control of the rate of the mixing of the fuel and combustion air, and consequent reaction times, in order to produce fuel rich zones which inhibit the formation of NOx from fuel nitrogen (RRIPG, undated).

These reductions are achieved by careful attention being paid to the geometry of the burner, and the use of ports to admit combustible air. The relative importance attached to either of these factors can depend upon the fuel selected. In the case of gas, for example, which usually has a nitrogen content of less than 0.1 per cent, fuel nitrogen NOx only accounts for a 20 per cent maximum of the total NOx produced and flame temperature is therefore critical, whereas in the case of coals having a fuel nitrogen content varying between 1 per cent and 1.5 per cent, fuel nitrogen NOx accounts for some 50-70 per cent of the total NOx produced (RRIPG, undated)

Low NOx burners can be located in the front wall of the boiler combustion chamber, or in the corner of the chamber to achieve a tangential firing system. Front wall low NOx coal burners can achieve a NOx reduction of 50 per cent or greater (depending on coal quality) compared with existing standard front wall coal burners, and can be retrofitted to existing boilers without modification to boiler pressure components (Allen, 1990). Tangential and wall fired systems, however, can achieve even greater NOx reductions with a combination of appropriate burner design and configuration within the combustion chamber, together with the use of particular operating systems which reinforce the selection of burner design and configuration to provide additional reductions of NOx emissions. These techniques include:

overfire air, to achieve staged combustion;
low NOx concentric firing systems, which can provide increased staging of combustion, and fuel rich combustion centre zones and air rich outer zones;
rich or lean burners, which can be used to extend the principle of fuel rich/air rich combustion to individual nozzles;

reburn, in which some of the fuel is separated from the main burner and re-injected into the combustion system.

These operating systems can be used in varying combinations to achieve different results as they are not necessarily additive in their individual effects, but when all of them are operated together with low NOx tangential or wall-fired burners, it is claimed that NOx reductions of 80 per cent can be obtained (Allen, 1990), compared to previous front wall burners. Practical reduction of 50-60 per cent from a baseline emission level of 1100-1400 mg/Nm^3, to levels below the EC LCPD level of 650mg/Nm^3 is the more frequently quoted practical statistic however (Allen, 1990; RRIPG, undated; Raymant, 1990; DoE, 1991, pp. 101-36). Low NOx combustion modifications are also apparently not expensive as they only cost a few pounds per kilowatt (between £5 and £15, 1990 prices) of installed electricity generating capacity (Raymant, 1990), within a total investment of up to £700-£800/kW (Boehmer-Christiansen and Skea, 1991, p. 34), or some £1.8-£10.0 per kWt annualised capital cost for existing boilers, and £0.4 per kWt for new boilers (1991 prices). In addition, the annual operating and maintenance costs are considered to be negligible for low NOx burners on new boilers and 0.0007-0.006p/kWh (1991 prices) on existing boilers, providing a figure of £14 per ton of NOx abated for new boilers, and £132 per ton for second generation low NOx burners (DoE, 1991, p. 135) in existing boilers. Low NOx burners are generally regarded as the best available technology in most of Europe and USA, although Germany and Japan have selected the post-combustion method of selective catalytic reduction for NOx reduction when using certain types of coals (IEA, 1988, p. 29). This method is described later in this section of the chapter.

The manufacture of low NOx burners is not particularly difficult, requiring mainly machining, welding and other fabrication processes, although care is required during assembly to achieve the requisite precision with other fabrications in the boiler, such as banks of hot water tubing and air ducting. From the technological viewpoint, therefore, there would appear to be no great difficulty in the transfer of manufacturing know-how for low NOx burners from the West to existing boiler factories in the former USSR, although design modifications may be necessary to suit the combustion characteristics of coal supplies in the former USSR, and development may also be required of quality management and supply chain procedures. Furthermore, a significant amount of pre-development work will be required to measure the performance of an existing boiler in its present regime; and the fuel properties, combustion chamber geometrics and fuel and firing air velocities, at existing installations.

The potential for the installation of low NOx burners is very high in the former USSR, because of the high levels of coal-fired electricity generation capacity in that region, particularly in the Urals and parts of Siberia, and the potential for Russian boiler factories to assimilate this proven Western technology. In view of the limited available investment resources in that region, however, it is clearly important to select priorities for any potential retrofit programme according to age and capacity of power stations, the necessity for major overhauls, the appropriateness of low NOx techniques for the types of coal being burnt, and the technical capability of Russian boiler factories to produce the necessary equipment. There is clearly a potential for the fitting of low NOx burners in new power stations, although new investment programmes are likely to be delayed in the current economic conditions of falling industrial output. This question of Russian prioritisation is discussed in Chapters 5 and 7 below.

There is further evidence to suggest a potential for the application of low NOx technologies in the former USSR as power engineering development establishments in that region are engaged in discussions with a Western manufacturer of this equipment (see Chapter 6)· and attempts have been made at domestic manufacture and installation. A verbal report on the operation of these installations reveal doubts on their levels of success, however, and the present author has yet to locate a report of any West/East commercial transfer agreement in this area of technology. Published reports in the Russian technical press appear to suggest a preference for developments in coal-fired 'post-combustion' rather than 'in-combustion' low NOx developments (Fillipov, 1990; Veshnyakov et al., 1990), and to limit the application of low NOx burners to gas and oil-fired boilers (Gribkov et al., 1993; Kotler and Enyakin, 1994). References to coal-fired low NOx burners are often in the context of Western developments (Rodgers and Morris, 1994) rather than indigeneous expertise which only appears capable of achieving reductions of some 15 per cent of NOx emissions from boilers which could be emitting up to 1200mg/m^3 if produced before 1989 (Fillipov, 1990). Such reductions do not meet the atmospheric emission requirements of *Goskompriroda* when burning fuels available from the former USSR, except for natural gas and brown coals which has consequently influenced a preference for post-combustion cleaning for the use of hard coals (Fillipov, 1990; Veshnyakov et al., 1990). The selection of these more expensive technologies, however, may not be the best use of scarce investment resources in the current economic climate in the former

Soviet region, and a more recent publication reports the development of low NOx combustion methods to reduce NOx emissions by 30 per cent, and some practical successes with the low NOx combustion of Berezovsk and Ekibastuz coals (Kotler and Enyakin, 1994).

In addition, the greater attention devoted to oil and gas burners is probably due to the fact that 70 per cent of thermal and electrical energy is produced in oil and gas-fired power stations and boilers (Kotler and Enyakin, 1994), (although gas is by far the single most widely used fuel, as discussed in Chapters 3 and 5). For boilers having steam capacities varying between 320 tons/hr and 3950 tons/hr, NOx reductions from $1500mg/m^3$ to $100mg/m^3$ have been achieved for gas; and reductions from $1320mg/m^3$ to $110mg/m^3$ for fuel oil, depending upon combustion conditions (Kotler and Enyakin, 1994).

Limestone Injection Multi-Stage Burners (LIMB)

Limestone Injection Multi-stage Burners (LIMB) are reported as a promising type of 'in-combustion' technology which simultaneously reduce SOx and NOx emissions, and also contribute to lower particulate emissions levels. NOx reductions are achieved through low NOx combustion techniques of improved burner geometry, and control of combustion temperature and air/fuel mixture; whilst SOx emissions are controlled through absorption into limestone at or near the burner. The NOx reductions are similar to those obtained from low NOx front wall burners, namely some 50 per cent, and the SO_2 emissions can be reduced by some 50-60 per cent (IEA, 1988, pp. 67, 68). Although these reductions may be acceptable from the NOx viewpoint, they may not be acceptable from the SO_2 reduction for new plant, unless used in conjunction with coal cleaning, although they could be acceptable for some types of retrofitted plant. At present, however, the LIMB process has yet to be proven on a commercial scale, and cannot at present be considered to be a viable emission reduction for technology for transfer to the former USSR, although there may be scope for using combustion development facilities in that region for further research and testing of LIMB burners.

Post-Combustion Systems

Introduction

Post-combustion systems for the control of pollutant emissions work on the principle of extraction of the relevant gases from the flue of the boiler. The two most widely used systems are flue gas desulphurisation (FGD) for the reduction of flue gas sulphur dioxide and selective catalytic reduction (SCR) for the reduction of NOx. These systems have the advantage of being highly effective in terms of pollutant extraction, but the disadvantage of high costs to install and operate.

Flue Gas Desulphurisation (FGD)

The most common form of FGD is the wet lime or limestone scrubber, in which a wet slurry of lime or limestone is sprayed through the flue gases. This compound reacts with the sulphur dioxide to form a calcium sulphite sludge plus carbon dioxide. The calcium sulphite can then be oxidized to form gypsum which may then be sold for use as an aggregate or for use in the manufacture of wallboard. Six per cent by weight of limestone is required for the coal tonnage burnt, and approximately ten per cent of the coal tonnage is produced as gypsum (Boehmer-Christiansen and Skea, 1991, p. 45).

Other alternatives to the limestone method include the Wellman Lord regenerative system, which recovers elemental sulphur or sulphuric acid; and SO_2 reduction by a process of oxidation. These systems are viewed as being less harmful to the environment than the limestone method since large quantities of sludge do not require disposal, but 75 per cent of installations use the limestone method because they are less expensive (Boehmer-Christiansen and Skea, 1991, p. 45), and they also have a comparatively high reliability (IEA/OECD, 1993, p. 10).

Although FGD systems are extremely effective, reducing SO_2 emissions by up to 90 per cent, they are also expensive to install. The German retrofit programme over 37,000MW(e) cost some 380DM/kW, whilst the British retrofit programme is expected to cost of the order of £150/kW (1991 prices). Furthermore, the operating costs are also high leading to cost increases of some 15-20 per cent per kW compared with conventional power stations. Nevertheless, FGD systems had been installed into more than 150,000 MW(e) by 1991 worldwide with some 90 per cent of these installations being located in the US, Japan and Germany.[1] In addition, the UK has recently embarked on an FGD programme of some 23,900MWt (Boehmer-Christiansen and Skea, 1991, p. 260), with some 4,000MW already commissioned at the Drax power station (*National Power Environmental Performance Review 1991*). It remains to be seen if these programmes are maintained, however, bearing their costs in mind, and the opportunities for conversion to cleaner fuels and newer combustion processes.

Russian specialists are well aware of the various options available for flue gas desulphurization, as research in this field was commenced prior to the Second World War, and an ammonia regenerative plant was built at Mosenergo TETs 12 (Combined Heat and Power [Co-generation] Plant No.12) power station during the 1950s which desulphurised some 200,000m³/hr of emission gases. In the early 1960s, a large limestone desulphurisation plant was then built at the sinter plant of the Magnitogorsk Metallurgical Combine, which handled some 3 million m³/hr of emission gases (Shirokov, 1990).

During the following twenty years, however, Soviet developments appeared to be slow compared with those of their Western counterparts. It was only in 1982 that a large unit (1 million m³/hr) for gas cleaning using the ammonia regenerative method was commenced at Dorogobuzhskii TETs, and it took some eight years to reach commissioning. In addition, an experimental limestone unit was commenced at Gubkinskii TETs, having a capacity of 100,000m³/hr of gas, and capable of producing high strength gypsum. Proposals were subsequently made to build some 160 gas cleaning units with total emissions of 200 million m³/hr (Shirokov, 1990). Wet limestone installations were planned at Zuevskii GRES (State Regional Electrical Power Station), Zmievskii GRES, Gusinoozersk TES (Thermal Power Station), Ulan-Udenskii TES, Ekibastuz GRES 2, Tselinograd TETs 2 and Kuznetskii TETs, together with a dry method experimental unit at Omsk TETs 5, and a dry/wet method experimental unit at Chelyabinsk TETs 3 (Fillipov, 1990). It is difficult to evaluate how far the plans have progressed, however.

This slow progress in the former USSR, however, is viewed by Russian authors as being in stark contrast to progress in the West, where tight and enforceable legislation, together with research and development on combustion and post-combustion processes have facilitated the implementation of flue gas desulphurisation systems. Furthermore, the original programme for Soviet flue gas desulphurisation envisaged the installation of cleaning units simultaneously with the boiler system (Shirokov, 1990), whereas some Western systems have been installed alongside existing thermal power stations.

Russian specialists consequently appear to be looking for opportunities for the assimilation of Western expertise in this field, but are also concerned about the costs of FGD systems (quoted as varying between 15 per cent and 20 per cent of the cost of the power station) (Veshnyakov et al., 1990), and the consequent effects on power generation costs (10-30 per cent increase) (Shirokov, 1990). Furthermore, the most-favoured system in the West (wet non-regenerative limestone system) fails to facilitate any increase in power station efficiency or decrease in sulphur dioxide generation during combustion, and requires additional transport infrastructure to deliver limestone to the power station and remove the gypsum. Estimates have been made that 1.7 tons of limestone are required to remove 1 ton of sulphur dioxide contained in emission gases, which then produces more than 3 tons of dual-acqueous gypsum. For a typical 200MW generation unit, burning Donetsk coal emitting some 1 million m^3/hr of fumes, some 245 tons of limestone will be required daily, producing some 460 tons daily of dual aqueous gypsum (Veshnyakov et al., 1990).

Selective Catalytic Reduction (SCR)

Selective Catalytic Reduction methods use a combination of ammonia and catalytic honeycombs or plates to reduce NOx in the flue gases, and up to 80-85 per cent reduction can be achieved. This is a higher proportion than most low NOx combustion processes, but the SCR system is more expensive to run as the catalysts need to be replaced every 4-5 years, and the installation costs in Germany were some DM210/kW spread over 33,000 MWe, and these high costs are also evident in other published sources. SCR is regarded as the best available technology in Japan and Germany, and German regulations require its installation in hard-coal fired stations but not for those fuelled by brown coal, oil or gas.[2]

Selective catalytic reduction methods appear to have been the preferred method to date for the reduction of NOx emissions in the former USSR, although some concern has been expressed about the disposal of the ammonia by-product in urban environments (Okhotin, 1992). Serious development work has only been carried out since 1986, however, and before that date emissions varied between 300 and 1200mg/m^3. As initial development work on primary combustion control could only achieve 15 per cent reductions from that level (Fillipov, 1990), most attention was subsequently devoted to SCR because of the possibility of achieving 80-90 per cent reductions. These methods were preferred to selective non-catalytic reduction because of difficulties in temperature control in the region of ammonia input for that process, although recent interest has also been shown in carbamide (urea) insertion (Veshnyakov et al., 1990). Development plans exist for the fitting of ammonia-catalytic methods to five power stations (Zuevsk GRES 2, Zmievsk GRES, Ekibastuz GRES 2, Tselinograd TETs 2, and Kuznetsk TETs) (Fillipov, 1990) but there is no published information available on the progress of those plans.

Although the installation of SCR systems has been found to be expensive in the West, it is possible that such costs will be lower in the former USSR, in view of the high labour content involved in these installations, and the comparatively low labour costs currently operating in the former USSR. Furthermore, policy-makers in that region may prefer those systems in view of their higher rates of emission reduction when compared with low NOx systems, but it is still not apparent that SCR is the most appropriate method of NOx reduction in that region with its current level of shortage of investment funds.

GAS COMBUSTION

Gas-Fired Steam Boilers

Natural gas combustion provides the cleanest alternative for electricity generation by thermal means, as this fuel has a negligible sulphur content, and its nitrogen molecular structure is conducive to low NOx combustion. Furthermore, the CO_2 emissions from this fuel are only some 58 per cent of those of hard coal per unit of heat generated (Schemenau and van den Berg, 1990). Most gas-fired power stations in the former USSR, but particularly those converted from the previous use of oil and coal have operated gas-fired boilers with steam turbo-generation as previously outlined in Chapter 3 (see Figure 3.1).

The Soviet state standard relating to NOx emissions from various types of boiler (GOST 28269-89) specified NOx gas boiler emission limits varying between 125 and 290mg/Nm^3 depending upon the steam capacity and age of the boiler, and Veshnyakov et al. (1990) quote NOx emission limits varying between 125 and 250mg/Nm^3 for boilers fitted with gas cleaning (see Chapter 2). Both of these sets of limits compare favourably with the EC LCPD NOx limit of 350mg/Nm^3 for gaseous fuels. Clearly further research is required on this topic as state-specified emission standards may not always be observed in practice, and data presented in a subsequent section of this chapter refers to NOx emissions of 403mg/Nm^3 from gas-fired steam boilers. There is also evidence to show that some large gas-fired steam boilers (between 950 tons/hr and 3950 tons/hr capacity) were emitting some 1200-1500mg/m^3 of NOx in the early 1990s, before modification to reduce those levels to 100-300mg/m^3 (Kotler and Enyakin, 1994), using staged combustion.

In addition, similar studies have been carried out by the Institute of High Temperatures of the Academy of Sciences in conjunction with the KEMA company of the Netherlands. As a consequence of modifications to the gas burners and fuel feed to 18 burners in $2 \times 300MW$ power sets at TETs 21 operated by Mosenergo and burners in a 300MW unit at the same power plant, the NOx emissions were reduced from $250mg/Nm^3$ to $70mg/Nm^3$ and $100mg/Nm^3$ respectively.

It appears, therefore, that sufficient expertise exists within the Russian Federation to reduce the levels of NOx emissions from gas fired boilers to those specified in the European Union's LCPD using low NOx burners, and any barriers to extensive retrofitting of gas fired power generation plant are therefore likely to be commercial rather than technical.

Gas Turbines

Gas turbine development in the former USSR, as in the West, has been carried out in four main areas of application, namely aero-engines, marine engines, stationary turbines for pumping units in gas pipelines, and stationary turbines for power generation.

Aero-engine applications have been at the vanguard of gas turbine development, as a consequence of the necessity for aero-engines to continually improve power to weight ratio, and reduce fuel consumption. These developments have required increases in combustion temperature to obtain more power from the thermodynamic cycle; and temperature increases, in their turn, have required consequent developments in combustion chamber and turbine blade design with associated use of lightweight heat-resistant materials. Western engine manufacturers have also paid high levels of attention to reduced levels of noise and NOx emissions to meet international standards. The developments thereby gained in aero-engine technology have been transferred to 'aero derivatives' of 25-40MW applications used in marine, gas compression and power generation applications, and also diffused to larger framed power generation turbines having capacities up to 250MW.

In aero-derivative gas pipeline pumping applications, natural gas tapped off from the pipeline is ignited in a gas generator derived from an aero-engine design, to provide a high velocity stream of hot exhaust gases to drive a gas turbine. The power-take-off shaft from this turbine is used to rotate the drive shaft of a compressor unit which compensates for pressure losses along the gas pipeline. This design enables the speed of the aero-engine/gas turbine/rotary compressor configuration to be rapidly modified in line with variations in pipeline pressure, and to thereby modify the output conditions accordingly. For power generation purposes, however, the gas compressor unit is replaced by an electricity turbo-generator (see Figure 4.1), and it is also sometimes necessary to introduce an additional gearbox. This extra unit is required as the steady output speed required from the turbo-generator set to achieve the required frequency of electric current, may be different from that achieved at the optimum operating thrust of the aero-engine. Aero-derivative gas generator configured turbines have the advantage of rapid response, which enables this type of configuration to respond rapidly to peak loading conditions.

The stationary 'frame-type' type of power generation gas turbine, however, is designed to achieve high torque and efficiency operation. Gas ignition occurs within its own combustion chamber, at the rated power and speed of the turbine, driving a common shaft for the compressor, turbine, and generator configuration (see Figure 4.1). In addition, power generation turbines are built in various frame sizes up to some 250MW in capacity, and this range of capacities combined with their efficiency characteristics cause them to be used in a far wider range of higher power generation applications than the aero-engine/gas turbo-generator configuration. In 'frame-type' power generation gas turbine applications, therefore, aero-engine technology is used to improve specific components such as blade designs and coatings rather than the implementation of complete aero-derivative configurations. The most recent aero-derivative developments, however, also use a configuration similar to the frame- type turbine, and offer advantages of higher efficiency. The choice of turbine therefore is influenced by total capacity required, gas prices, relative efficiencies, and the comparative costs of several aero-derivative units compared with one frame-type turbine.

These similar paths of development appear to have been followed in both the former USSR and the West, although Western developments appear to have occurred at a faster pace. Furthermore, although turbine manufacturing establishments in both the former USSR and the West could be divided between those in the aero-engine sector and those in power engineering, Western developments between the two sectors have been facilitated by cross-ownership between the two sets of companies, or the sharing of technical data required for product development. In the former USSR, however, no such cross-ownership has existed, and both sets of manufacturers were responsible to separate industrial ministries, either the Ministry of Aviation Industry (*Minaviaprom*) in the military sector; or the Ministry of Power Engineering (*Minenergomash*) or Ministry of Heavy Engineering (*Mintyazhmash*), in the civilian sector. The transfer of technical data between establishments in these different industrial ministries is likely to have been limited in the former USSR, and especially between the military and civilian sectors for reasons of national security.

In addition, the ratio between peak and base loads have been higher in the West than in the former USSR as a consequence of a higher proportion of Western electricity output being consumed by households (27 per cent in Western Europe compared to 8 per cent in the former USSR, according to IEA, 1994, p. 197). The higher variability of electricity demand in the West is therefore likely to have promoted the development of the gas turbines required to meet these peak requirements.

According to data presented in Soviet technical journals, the efficiencies of gas turbines manufactured in the former USSR have lagged behind those of their Western counterparts, and differences in efficiencies appear to have been caused mainly by higher levels of gas inlet temperatures in the Western machines. A high inlet temperature is also of particular relevance to efficiency of combined cycle systems (see below), where exhaust gases of some 500-600°C are required to produce steam for use in a steam turbine, following a gas temperature drop of some 200-300°C between gas turbine inlet and outlet. Table 4.1 below shows data provided for the LMZ range of power generation gas turbines produced at the Leningrad Metal Works, the former Soviet Union's largest manufacturer of steam, hydro- and gas turbines for power generation (see Chapter 6 below). It is important to note from the published data that although the inlet temperature for one of the turbines provided in Table 4.1 is for an obsolete model (GT-100), the proposed updated version of this machine (GT-140-1100) will have an inlet temperature approaching that of the cited ABB machine, originally produced in 1987, but more than 100°C lower than the Siemens machine originally produced in 1992. Identical information is also provided in that table on a 140MW modified version of the GT100 following development work at a Mosenergo GRES for peak loading applications. The data shows increased inlet temperatures and efficiency for the modified turbine, and the authors of that paper (Osyka and Efimov, 1994) suggest that scope for similar modifications may exist for two turbine sets at Krasnodar TETs, another two sets at Simferopol TETs, and three sets at Ivanovskaya GRES as well as to the three turbine sets installed in 1978 and 1980 at the Mosenergo GRES.

In addition to differences in efficiencies, there are also differences in the levels of NOx emissions between gas turbines manufactured in the former USSR and their Western counterparts. Carette and McMillan (1990) quote NOx emissions of 25ppm or 51mg/Nm^3 (at 15 per cent O_2) for both dry (at

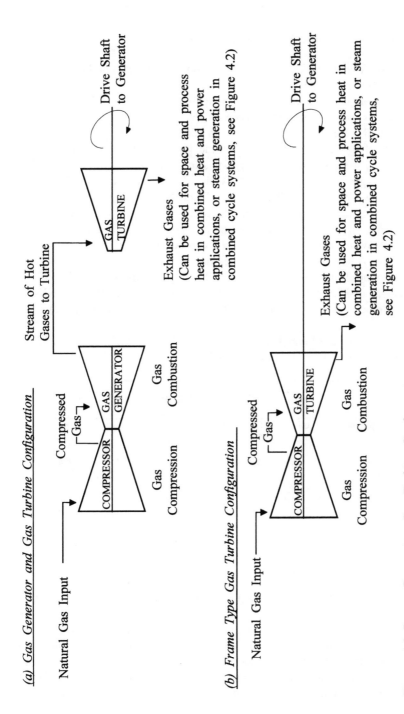

(a) Gas Generator and Gas Turbine Configuration

Natural Gas Input

COMPRESSOR

GAS GENERATOR

Gas Compression

Gas Combustion

Compressed Gas

Stream of Hot Gases to Turbine

GAS TURBINE

Drive Shaft to Generator

Exhaust Gases
(Can be used for space and process heat in combined heat and power applications, or steam generation in combined cycle systems, see Figure 4.2)

(b) Frame Type Gas Turbine Configuration

Natural Gas Input

COMPRESSOR

GAS TURBINE

Gas Compression

Gas Combustion

Compressed Gas

Drive Shaft to Generator

Exhaust Gases
(Can be used for space and process heat in combined heat and power applications, or steam generation in combined cycle systems, see Figure 4.2)

Figure 4.1 Power Generation Gas Turbine Configurations

combustion temperatures over 1100°C) and wet (100 per cent down to 75 per cent load) low NOx techniques.

An option which is repeatedly discussed in the Russian technical press to improve power station performance, is the conversion of aero-engine gas turbines to civilian power generation, in addition to their application for the driving of gas compressors and marine power units. One of the most commonly used turbines in gas transmission applications has been the NK series of gas turbines developed by the *Trud* Scientific and Production Association in Samara. The NK 12 ST and NK-16ST were developed for gas transmission in the 1980s, and the Russian state gas company *Gazprom* has some 1169 of these derivatives in operation, accounting for some 11,500MW in total or some 32 per cent of installed gas transmission capacity. A new gas pipeline compressor turbine unit (NK 36ST) was due to be introduced in 1993, capable of achieving a capacity of 25MW and efficiency of 36.4 per cent, and a stationary power station turbine (NK-37) was also being developed on the basis of that technology (Orlov, 1992).

Table 4.1   Comparative Data on Power Gas Turbine Efficiencies and Inlet Temperatures (LMZ and Western Models)

Firm	LMZ[1]	LMZ[1]	LMZ[1]	ABB[1]	Siemens[1]	LMZ[2]
Type of turbine	GT-100	GTE-150	GT-140-1100	13E	V 94.3	Modernized GT100
Year of 1st Production	1968	1990	future design	1987	1992	1994
Capacity MW	100	157.6	140	148.8	200	140
Efficiency %	28.0	31.0	36.0	34.8	35.4	36
Inlet temperature°C	750	1100	1100	1115	1260	1100

Sources: 1. Polishchuk (1993).
      2. Osyka and Efimov (1994).

One Russian writer (Polishchuk, 1993) sounds a note of caution on the uncritical development of such a programme, however, by pointing out that although the GE LM 6000 40MW turbine achieves efficiencies of more than 40 per cent and can be considered as a converted (or aero-derivative turbine), such power sets can usually only operate from a narrow fuel base (usually natural gas and light fuel oil which may not suit Russian operating conditions), and the cost of conversion from military to civilian application can sometimes exceed the costs of designing a special purpose stationary turbine. Polishchuk continues by pointing out that the capacity of aero-derivative engines does not usually exceed 50MW, although a new 110MW turbine (the GE LM 8000) to be designed to achieve an efficiency of 47 per cent and an inlet temperature of 1360°C uses many features derived from aero-engine technology. Comparative technical data for a range of both aero-derivative and electrical power engineering turbines are provided in Table 4.2, together with information on their location of manufacture.

It is apparent from these data, therefore, that factories and design establishments in the former USSR have extensive design and manufacturing expertise in gas turbine production, particularly in the area of aero-engine derivatives for gas transmission applications. It is likely that this expertise will continue to be demanded for pipeline applications, in view of the importance of natural gas for both domestic and international markets, and the necessity of equipment replacement for many of the gas pumping stations installed more than twenty years ago. This replacement programme will probably be met by both imported and domestically produced equipment, and using turbines produced through joint production arrangements with Western companies. These latter arrangements will have the advantage of combining Western design expertise with manufacturing capacity availability in the former USSR. Furthermore, although some gas turbines produced in the former USSR have been comparatively advanced in terms of performance parameters and design concepts, Western-designed machines are reputed to have higher levels of reliability although this parameter is clearly influenced by working environment and servicing programmes as well as machine design and manufacturing conformance.

In terms of electric power generation, however, it is likely that increased use of gas turbines will be in the area of combined cycle operation, because of the increased efficiencies and total output available from these generating systems, compared with gas turbine installations alone. Combined cycle systems are described in the next section of this chapter, together with further discussion on the levels of atmospheric emissions.

Combined Cycle Systems
The most simple form of combined cycle system consists of a turbine powered by the combustion of natural gas or oil, supplemented by a steam turbine. The steam for this latter turbine is generated in a heat recovery steam generator heated by exhaust gases from the gas turbine (usually some 500-600°C), or in a heat exchange unit which combines these exhaust gases with gas, oil or coal firing (see Figure 4.2). Combined cycle systems exhibit

Table 4.2 *Comparative Data on Gas Turbine Efficiencies and Inlet Temperatures (Aero-derivative and Power Generation Turbines)*

Model	89ST[1]	D-18[1]	R-29-300[1]	NK-37[1,2]	RD-36-51[1]	GTE-25[2]	GTE-45u[2]
Manufacturer				KMZ[2,4]	RMZ[4]	TMZ[4]	
Year				1994[2]		1993	
Nominal Capacity (MW)	22.5	25	26	30[1];25[2]	61.7	25.5	40.2
Efficiency %	30	34	30	38[1] 35.7[2]	33	32.3	33.8
Inlet Temperature °C	1000	1217	1092	1147[1]	1019	1060	1230

Model	GTE-30[2]	GTG-15[2]	GTG-110[2]	GTE-45[2]	GTE-115[2]	GTE-200[2]		GTD-110[3]	
Manufacturer	NZL[4]	YuTZ[4]		KhTZ[4]		LMZ[4]		Zarya[4]	
Year	1990	1990	1994	1989	1993	1989	1993	proposed	1992
Nominal Capacity (MW)	30	15	110	54	118.9	128	157	185	110
Efficiency %	29.0	34.5	36.0	28	34.2	30.5	31.0	32.6	36.0
Inlet Temperature °C	900	900	1210	900	1170	950	1100	1250	1240

Notes to Table 4.2

1. The data for the aero-derivative engines (89ST, D-18, R-29-300, NK37 and RD-36-51) are taken from Aminov et al. (1994).
2. The data for the stationary power turbines (GTE-25, GTE-45u, GTE-30, GTG-15, GTG110, GTG-45, GTE-115, GTE-150 and GTE-200) are taken from 'Perspektivy primeneniya gazovykh turbin v energetike' (no authors cited), *Teploenergetika*, 1992, No. 9, pp. 2-8. This source also cited a nominal capacity of 25MW for the NK-37, and an efficiency of 35.7%. It should be noted that this paper, published in 1992 refers to performance specifications in 1993 and 1994 as well as those in 1989 and 1990. The specifications for 1993 and 1994 presumably relate to anticipated parameters for those turbines. Additional data for the NK-37 aero-derivatives turbine is also provided in Orlov (1992). The capacity is cited as 25MW, the efficiency as 36.4%, and the inlet gas temperature as 1147°C (1420°K).

3. The data for the GTD-110 turbine has been obtained from Romanov et al. (1992). These authors also provide peak performance data in addition to the base data shown in the above table. The performance data is for use with natural gas or aero-engine fuel.

4. The author has been unable to locate the full name of the factory for which *KMZ* is an abbreviation, but the NK-37 was apparently developed (and probably manufactured) at the '*Trud*' Scientific and Production Association in Samara (*SNPO*) (see Orlov 1992; Polishchuk 1993). *KMZ* may therefore be an abbreviation for Kuibyshev Aero-Engine Factory (*Kuibishevskii Motorstroitel'nyi Zavod*) in Samara. Furthermore, the author has been unable to locate the full names of the factories for which *TMZ*, *NZL* and *YuTZ* are abbreviations, although *TMZ* may be in Tula (*Tul'skii Motorstroitel'nii Zavod*) and *NZL* may be an abbreviation for the *Nevskyi Zavod Leningrad*. *KhTZ* is an abbreviation for the Khar'kov turbine Factory (*Khar'kovskii Turbinyi Zavod*) (Polishchuk 1993). *KhTZ* ('Perspektivy...', 1992) and *LMZ* is an abbreviation for the Leningrad Metals Factory (*Leningradskoe Metallicheskii Zavod*) (Polishchuk 1993). *KhTZ* is also sometimes referred to as *Turboatom*. The *Zarya* factory is in Nikolaev, and is producing the GTD-110 to NPO *Mashproekt* designs (Romanov et al., 1992). Polishchuk (1993) also refers to the *Saturn* (or A.M. Lul'ka) Scientific and Production Association which has designed the AL-31F aero-derivative for power generation, and this turbine is considered to be analogous to the GE LM-8000. The RD-36-51 is manufactured by the Rybinsk Aero-engine Factory (*Rybinskoi Motorostroitel'nyi Zavod - RMZ*) for installation on to the TU144 aircraft. This engine has a compressor input airflow of 273 kg/sec, with potential for upgrading for 250-300 MW installations (see Batenin et al., 1993) Reports sometimes refer to the *UMTZ* factory (which is probably the Urals Aero-Engine Factory (*Ural'skii Motorno-Turbinyi Zavod*) or the Ufa Aero-Engine Factory (*Ufimskii Motorno-Turbinyi Zavod*), but the author has not been able to find any information relating to technical specifications for turbines produced at this factory. *YuTZ* may be an abbreviation for the Southern Turbine Factory (*Yuzhnii Turbinnyi Zavod*), which might be located at or near to the *Zarya* factory at Nikolaev in Southern Ukraine. Both *YuTZ* and *Zarya* produce 110MW gas turbines (GTG-110 and GTD-110, respectively), but a further confusion is caused by the reference in another Russian source to the GTG-110 being produced to *Mashproekt* designs in the Nikolaev factory. Although the nominal power is the same as that given in Table 4.2 above, the year of production is given as 1993, the efficiency at 25% and the inlet temperature as 1220°C. (See Osyka and Efimov, 1994). A possible explanation may be that *YuTZ* is the factory making civilian power products at Nikolaev, whilst *Zarya* produce aero-engines.

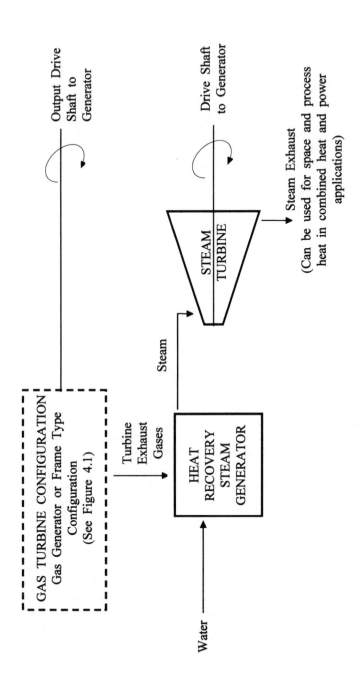

Figure 4.2 Combined Cycle System

the dual advantages of lower emissions per unit of power generated as a consequence of the substitution of gas for coal, together with high efficiency through the secondary use of the gas exhausts (Dibelius, 1988). Siemens (1994), for example, quote a base load efficiency of 36.2 per cent for a gas turbine terminal output of 222MW, compared with an efficiency of 53.2 per cent for a combined cycle of gas turbine and exhaust gas heated steam at a net block output of 223MW. If additional fuels are used to heat the steam, output can be increased although efficiencies may fall by some 6-7 per cent compared with the use of exhaust gases alone.

Combined cycle systems have the advantage of a shorter construction period and lower capital cost compared with new environmentally acceptable coal-fired power stations, and they can also be operated efficiently and cheaply at lower capacities than those demanded by coal-fired economies of scale. Furthermore, combined cycle power stations can provide an acceptable rate of return over a shorter time interval than their coal-fired counterparts, thereby making them more attractive to Western private investors at current fuel prices. These advantages for the former USSR need to be seen in the context of possible changes in fuel prices, however, and government policy decisions on natural gas as a 'premium fuel' which can be used more efficiently in installations other than power generation (Heppenstall and Trimm, 1990).

At present Western fuel pricing levels, however, combined cycle systems have a clear economic advantage over new coal-fired power stations, although not necessarily over a fifteen year time period for existing power stations fitted with FGD equipment (Newbery, 1993). To a certain extent, therefore, the advantages to be gained from the use of combined cycle equipment will depend upon the age profile of existing power stations and the associated degree of wear and tear to existing boilers and steam turbines, the levels and relative changes in the price of fuels, and the time interval over which an appropriate rate of return is required. The age profile of the power station and serviceability of the equipment will determine the extent to which existing steam-powered plant can be 'repowered' or 'topped up' with gas turbines to provide additional power, and heat recovery steam generators to provide the necessary volume of steam for the remaining steam turbines. In such installations, the previously installed steam turbines usually account for some 40 per cent of the total power generated, with the new gas turbines accounting for some 60 per cent. With increases in station efficiency from previous levels of some 30 per cent to almost 50 per cent, repowering can more than double the previous station output.

The economic implementation of such an option usually depends, however, upon the future serviceability of the steam turbines and the practical options for refurbishing and upgrading certain components, and the minimization of thermal stresses during repowered operation; and the necessity of replacement of the previous boiler. In general, however, steam turbines and their associated condensers have a longer service life than their boilers, and the physical size of gas turbine and associated heat recovery steam generating plant approximates to that of steam boilers, making boiler replacement a physical and cost-effective possibility. Alternatively, new steam turbines and heat recovery steam generators may be added on to existing gas turbine plant to provide some 67 per cent increase in output (Joyce, 1993).

In addition to the apparent commercial advantages of combined cycle equipment under conditions of privatization, gas turbine technology also has the overwhelming advantage of negligible SOx emissions (and consequently no need for expensive FGD equipment and subsequent disposal of large volumes of gypsum) and low NOx emissions (Raymant, 1990; Carette and McMillan, 1990). In the former USSR, however, it appears that gas turbines have been operating at NOx emissions of $180mg/m^3$ using natural gas, and $350mg/m^3$ using fuel oil, although it is intended to reduce these levels to 100 and $200mg/m^3$ respectively ('Perspektivy...', 1992). For aero-derivative engines, NOx emissions were quoted at $160-270mg/m^3$ (Aminov et al., 1994). It is apparent that these emissions are within the EC LCPD specification of $350 \ mg/Nm^3$ for gas combustion (although that limit is for boilers), but higher than levels claimed for advanced Western machines and sometimes exceeding the GOST 21199-82 limit of $220mg/m^3$ (Orlov 1992).

There is consequently scope for the transfer of Western technologies in the field of NOx emission reduction, as well as other fields of gas turbine technology, such as combustion chamber and blade design, and blade tip cooling. A recent article in the Russian technical press compares Russian and international expertise in the use of steam insertion in the combustion chamber to reduce NOx emissions and increase power output (Batenin et al., 1993), although the use of steam injection may not be the preferred option in the long term as it can lead to efficiency loss (ABB, ABB EV, undated).

There would consequently appear to be scope for the transfer of Western technology in the field of dry low NOx gas turbine combustion systems, particularly as there are verbal reports of local pressures in some Russian urban and environmentally sensitive areas to specify NOx emissions of 50 mg/Nm^3 for new gas turbine installations. When improved efficiencies are taken into account, combined cycle systems also have a distinct advantage of reduced NOx emissions (per unit of power generated) over gas turbines alone, which in their turn have an advantage over gas fired steam boilers (see Table 4.3) as a consequence of improved combustion.

According to Batenin et al. (1993) combined cycle systems can achieve the environmental advantages listed in Table 4.3, although it is not clear from the cited source as to whether the data is based on Russian or Western practical experience. Other data reported by Prutkovskii et al (1990) is provided in Table 4.4 and relates to performances from various combined cycle configurations, but it is unclear as to whether the data is based on Russian practical applications, reports from Western companies, or theoretical estimates.

Power generation industry personnel in the former USSR have gained experience in the utilisation of combined cycle systems, using a high pressure steam generator of 200MW capacity at the Nevinnomysskaya State Power Station which has operated for 83,000 hours on gas and liquid fuel; and two energy sets with low pressure steam generators of 250MW capacity at the Moldavian State Power Station, operating for 60,000 hours burning heavy fuel oil in the boiler and liquefied gas in the turbine combustion chamber. The energy set at Nevinnomysskaya Power Station obtained an 8.5 per cent saving in fuel consumption compared with steam turbine sets, and combined cycle sets with that configuration achieved a 25 per cent lower level of NOx (Prutkovskii et al., 1990).

There would appear to be little doubt over the capability of factories in the former USSR to design and manufacture steam turbines of the requisite performance to operate within a 'combined cycle' configuration, in view of their many years of experience in the design and manufacture of this equipment, although there may be some scope for the transfer of technologies relating to efficiency improvement and automation for steam turbines of this type. LMZ turbines, for example, are to be installed alongside Western-designed gas turbines in four combined cycle power stations in the former USSR, and also in international markets (see Chapter 6 below).

Table 4.3 Comparative NOx Emissions for Gas-fired Boilers, Simple Cycle Gas Turbines and Combined Cycle Systems

Indicator	Gas Fired Boilers	Simple-Cycle Gas Turbines	Combined Cycle Systems
NOx ppm	250	50	50
Efficiency %	39%	35%	55%
$\propto$	1.1	3.0	1.4
NOx mg/Nm^3, dry gas	513	102	102
Emission index, g/kg of fuel	7.9	4.2	2
NOx mg/Nm^3, $\propto = 1.4$, dry gas	403	218	102
NOx g/kWh	1.44	0.84	0.26

Source: Batenin et al., (1993).

Table 4.4 Specifications and Guarantees for Soviet Combined Cycle Systems

Indicator	Gas Turbine Configuration					
	2 × GTE-815			3 × GTE-85		
Ambient Air temperature °C Specification: Capacity (MW)	-15	0	15	-15	0	5
Gas turbine unit	2 × 103	2 × 91	2 × 78	3 × 103	3 × 91	3 × 78
Steam Turbine	93	90	86	141	135	129
Combined Cycle Unit (net)	292	264	36	440	396	354
Combined Cycle Efficiency (% net) Guarantees:	46.54	46.16	45.08	47.08	46.66	46.58
Capacity, net (MW)	270	243	217	406	364	325
Efficiency net (%)						
at full load	43.7	43.2	42.0	44.2	43.6	42.5
8 hours at 60% of nominal load	42.0	41.4	40.3	43.6	42.5	42.1
8 hours at 50% of nominal load	43.1	42.4	41.7	47.8	41.7	41.4
8 hours at 30% of nominal load	43.0	40.7	39.9	43.4	42.2	41.8

Source: Prutkovskii et al. (1990).

As heat recovery steam generators are major units in combined cycle systems, there is also scope for further research on these types of equipment, and the scope for technology transfer from Western companies. A paper published in 1992 suggested that there had been no installation of heat recovery steam generators in the former USSR at that time, as the Nevinnomysskaya GRES used sets of 35MW gas turbines with high pressure steam generators, and the Yakutskaya GRES used gas turbine exhausts for district heating. The authors of the 1992 paper suggested that further development work was required to improve the layout, steam parameters and corrosion resistance of heat recovery steam generators, in order to implement 350MW combined cycle systems based on two GTE115 gas turbines, and one 123MW steam turbine designed by the 'Turboatom' Association (formerly KhTZ) (Zarubin et al. 1992).

Other papers published in 1992, however, refer to a heat recovery steam generator manufactured by the Podol'sk Engineering Factory (formerly the 'Ordzhonikidze' Factory, Podol'sk) as part of a 130MW combined cycle package using two modified GTE-45U aero-derivative gas turbines from the TMZ factory, and a 40MW steam turbine produced by the Urals Heavy Engineering Factory (*Ural'skii Zavod Tyazhelogo Mashinostroeniya* [UZTM]) at Ekaterinburg (formerly Sverdlovsk). The development of the system was apparently contingent on the development of a prototype gas turbine in 1994, testing and a second prototype in 1995, with batch production to be commenced in 1996 (Akimov et al., 1992). There is also reference to heat recovery steam generators to be built at the Podol'sk Engineering Factory as part of a 325MW package to be installed in the Konakovskii GRES in 1996. This package will include two GTD-110 gas turbines to be produced by the *Zarya* Factory (Nikolaev) and one 100MW steam turbine from LMZ (Romanov et al., 1992.

Combined cycle systems may also be fuelled by gas produced from coal rather than natural gas, and the major features of this technology are described in the next section of this chapter.

ADVANCED COAL TECHNOLOGIES

Atmospheric Fluidised Bed Combustion

Fluidised bed combustion occurs below the atmospheric nitrogen fixation temperature of 900°C (compared to 1500°C for pulverized coal boilers) producing NOx levels within the range of 270-680mg/Nm^3 (Sondreal,

1992), which compares favourably at the lower levels with the current LCPD limit of 650mg/Nm^3; and another estimate provides a range even lower than this threshold value, namely 300-450mg/Nm^3 (Heppenstall and Trimm, 1990). Furthermore, the addition of limestone to the bed can also absorb up to 90 per cent of the total sulphur providing that the alkali to sulphur ratio is maintained between acceptable limits (Heppenstall and Trimm, 1990; Sondreal, 1992).

The two basic types of atmospheric fluidised-bed combustors (AFBC) differ in respect to fluidised gas velocity and mean particle size. 'Bubbling beds', with fluidising velocities between 1 to 2.5 metres per second and a mean particle size of 2-4mm, have been built to a capacity of 160MW; whilst 'circulating beds' have a gas velocity of 5 to 6 metres per second and mean particle sizes varying between 0.1mm to 0.3mm. Heat is extracted from the combustor and from a waste heat boiler for cooling the combustion gases before final clean-up, and superheated steam raised in the modified boiler systems drives a conventional condensing steam turbine. Circulating beds (CFB) appear to be gaining in popularity over bubbling beds as they can achieve improved combustion efficiency, sulphur removal and heat transfer, and capacity can be increased up to 500MW, although they appear to be favoured within the 100-250MWe range because of their flexibility (Heppenstall and Trimm, 1990; Sondreal, 1992).

Furthermore, CFB systems can be modified for use in 'repowering' or 'topping up' systems similar to the Pressurised Fluidised Bed (PFBC) plants described in the next section. AFBC systems offer a potentially economic method of repowering pulverised coal power plants to reduce atmospheric emissions, and estimates have been made that retrofitting can be achieved within a range of US$500-US$1000 per kW (1988 prices) (Sondreal, 1992). This range is lower than estimates of between US$1250 and US$1500 per kW for new pulverised coal boilers with FGD (Sondreal et al., 1994), and can therefore be considered as a viable alternative for future technology transfer to the former USSR, although the costs of operation and the processes of bed erosion may require further investigation.

Fluidised bed systems are being used in an increasing number of applications, but their commercial viability has yet to be demonstrated on the same scale as gas-fired and pulverised coal-fired systems. In the US, fluidised bed systems are in operation at the 160MW Shawnee (Kentucky) power station burning bituminous coal, the 125MW Black Dog power station in Minnesota, the 80MW Heskett power station in North Dakota, and the 110MW Nucla circulating bed plant in Colorado, and further data is being compiled from the use of these systems (Sondreal, 1992). Particular care is required over the investigation of fuel properties to obtain the optimal performance from AFBC systems, however, which will need to be considered in the context of the wide range of solid fuels used in power generation in the former USSR.

It is particularly important to bear in mind, however, that natural gas-fired combined cycle systems remain the cheapest stations to construct in terms of investment per unit of capacity (US$500-US$600 per kW) operating at efficiencies approaching 50 per cent, although using fuel at approximately double the cost of coal (Sondreal et al., 1994); and it is considered that advanced coal technologies will need to be reduced to US$800/kW to be competitive with combined cycle units burning natural gas delivered for US$ 3/MM Btu (Pitrolo and Bechtel, 1988; quoted by Sondreal et al., 1994). It is therefore important to determine the extent to which costs and benefits will vary between coal and natural gas firing in the former USSR. Labour costs for both scientific and skilled manual labour for the development and construction of CFBs, will need to be compared to the development work required for gas turbines and heat recovery steam generators to be used in combined cycle systems. In those regions which have used comparatively high levels of coal but have limited supplies of natural gas, fluidised beds could prove to be a viable option for power generation.

Pressurised Fluidised Bed Combustion (PFBC)

Pressurised Fluidised Bed Combustion (PFBC) is achieved by adding crushed coal to a bed of ash particles which have been fluidised by means of compressed air, typically at 10-15 bar and 850°C (Dawes et al., 1990). Particulates are removed from the combustion gases which are then expanded through a gas turbine. The exhaust gas turbine gases are used to pre-heat water which is then passed through tubes in the fluidised bed, thereby generating steam to drive a steam turbine in a combined cycle system (see Figure 4.3). NOx emissions are reduced compared with conventional systems as the temperatures of combustion are lower, and SOx emissions are reduced through the addition of limestone to the bed.

Furthermore, it is claimed that the thermodynamic efficiencies of PFBC systems (about 40.5 per cent) are higher than those of conventional pulverised fuel systems with flue gas desulphurisation (37-37.5 per cent). Experimental work has progressed on reducing corrosion problems, although the PFBC system has still to be demonstrated widely in practice to obtain detailed data on efficiency parameters (Dawes, 1990).

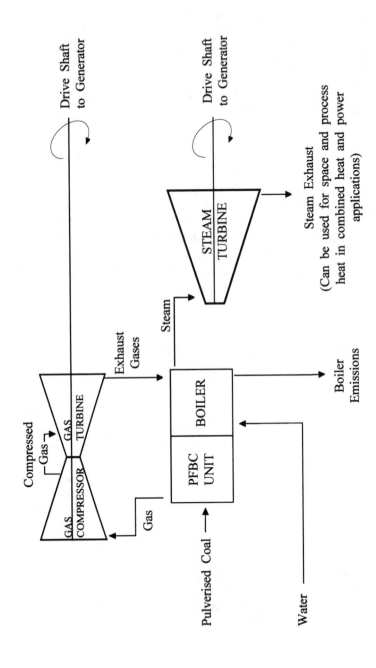

Figure 4.3 Pressurised Fluidised Bed Combustion (PFBC)

An ABB system comprising a PFBC turbine connected to a pressurised fluidised bed boiler system recently installed for American Electric Power (AEP) in Tidd, Ohio achieved a net efficiency of 35 per cent HHV using the existing 70MWe net output steam turbine, with NOx emissions of 150 mg/MJ (or 0.5lb/MMBtu [700mg/m^3]) and 90 per cent removal of sulphur emissions from 3.4 per cent S coal (or 0.51 lb/MMBtu [700mg/m^3]) (ABB, *PFBC...,* undated). Commercial systems are also being installed at Stockholm and Escatron (Spain), and a 330MWe scheme, partly funded by the US DoE is planned at the Philip Sporn Plant, West Virginia (Redman, 1989, quoted by Dawes et al., 1990). The capital cost of the Sporn demonstrator project was estimated to be some $2000 per kW, whilst that of the Tidd project was estimated at some $2393 per kW. These costs are higher than the $1700/kW level for pulverised coal-fired plants with FGD; but costs for mature PFBC combined cycle plants may be as low as $1560 per kW, and therefore competitive with pulverised coal firing (Sondreal, 1992). These estimates need to be proven under commercial conditions, however, although the Tidd plant was a re-powering demonstrator project using an existing steam turbine which was twice the size required for an optimal PFBC plant (ABB, undated).

Development work has also been carried out by British Coal on a PFBC topping cycle, using a PFBC combustor with an air blown PFBC gasifier to supply heat and gas to steam and gas turbine generating sets. This cycle should achieve an efficiency of about 45 per cent from gas turbines capable of operating at inlet temperatures of 1260°C. The efficiency increases compare favourably with the 38 per cent typically achieved from conventional coal-fired plant. Furthermore, there may be scope for wider use of this technology with high moisture coals, typical of those used in many parts of the former USSR (Sondreal et al., 1994).

The relevance of this technology to the former USSR depends, therefore, as in the case of AFBC systems discussed previously, on the relative costs of fuels, investment and future development work required. The US Clean Technology Programme, however, continues to support a number of advanced combustion projects in the US, and also in Eastern Europe, and it is therefore likely that further commercial information will emerge from these programmes which can promote the transfer of PFBC technologies to the former USSR.

Integrated Gasification Combined Cycle (IGCC)

As an alternative to the use of natural gas as a fuel, a combined cycle system may be coal-driven with the coal heated to obtain coal gas in an integrated gasifier. The coal gas thereby produced is used to fuel the gas turbine, and to also supplement the gas turbine exhaust gases during the heating of the water to drive the steam turbine (Dibelius, 1988) (see Figure 4.4). A proposed application of an Integrated Gas Combined Cycle (IGCC) system in Poland is considered to provide significantly reduced atmospheric emissions (11g/GJ of SOx and 52g/GJ of NOx) and improved efficiency (41.5 per cent) compared with a coal-fired steam turbine plant using untreated coal (1217g/GJ of SOx and 227g/GJ of NOx); whilst the use of cleaned coal in the gasification system is considered to provide a higher thermal efficiency (42.3 per cent) as a result of enrichment during the fuel cleaning process (Oudhuis et al. 1993).

In view of its demonstrated capability using East European solid fuels, there is scope for the transfer of IGCC technology to the former USSR, using combustion development establishments in that region to carry out some of the requisite research and testing. However, before commercial applications be considered on a wide scale, it is apparent that developments are required to reduce turbine component corrosion, with associated problems of down time. These problems require advances in gas purification to reduce the corrosiveness of included components (particularly retained alkalis), together with advances in turbine component technology to increase their corrosion resistance.

The reduction of retained alkalis has proved difficult to achieve without a reduction in temperature and consequent efficiency (Haupt et al., 1995). There is also still some concern amongst Western companies, that reductions in plant efficiency might occur as a result of excessive vibration through over-reactive surging response to variations in gas temperature. These concerns clearly need to be addressed before recommending the widespread transfer of this technology to the former USSR. Furthermore, the reported potential of Russian indigenous technological developments in the field of IGCC (Prutkovskii and Chavchanidze, 1995) may need to be evaluated from the viewpoint of turbine corrosion.

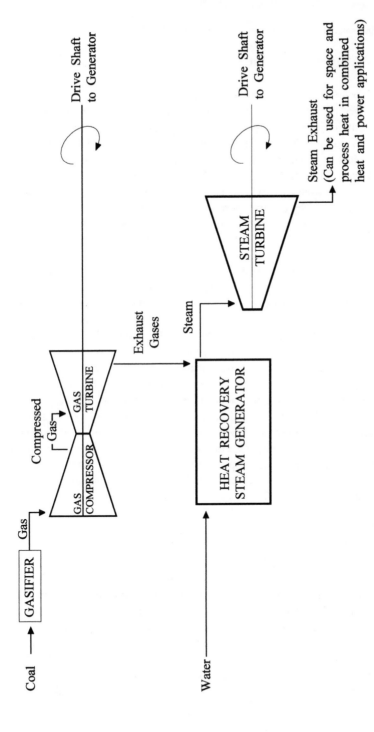

Figure 4.4 Integrated Gasification Combined Cycle (IGCC)

OPERATIONAL EFFICIENCY

A further option available for the reduction of atmospheric emissions during the power generation cycle is by means of improved power station operational efficiency. This option has the advantage that the savings lead to direct costs reduction through savings in fuel, whilst the levels of investments required are usually not high as the initial improvements will be in power station procedures, and the purchase of monitoring and control systems. Investments may also be made in insulation materials to reduce heat loss and thereby achieve increased efficiency.

In the Federal Republic of Germany, fuel consumption per unit of electricity fell by 75 per cent in the early 1990s, providing consequent reductions in SOx, NOx and CO_2 emissions. Similarly, reductions in emissions have been achieved as a consequence of American power station efficiency increasing to a current average of 33 per cent and most boiler-based generating systems provided with emissions control equipment have an overall efficiency of about 35 per cent, with the best plants approaching efficiencies of 40 per cent (DoE, 1991, p. 12). Scale efficiences have also been obtained in power generation in the former USSR since the 1950s, but it is likely that the potential for further scale efficiencies has now been exhausted. Gas-fired combined cycle systems now provide the opportunity for even higher levels of efficiency, and it is clear that opportunities exist for the transfer of relevant operational and combustion procedures and technologies to the former USSR, because of the many years of power engineering and electricity generation in that region. The scope for the implementation of such developments in practices, however, depends upon the existence of political, economic and organizational stability, and these are discussed in Chapter 7.

ENERGY SAVING

The demand for electric power can be significantly reduced by paying more attention to the methods by which energy is utilised, particularly in industrial applications, and specifically within energy intensive industries such as ferrous and non-ferrous metals production. Western sources cited in Russian journals claim energy savings of 20 per cent (Sitas and Streker, 1994), and such savings in energy consumption will have a direct effect on emissions from the power generation sector, as lower quantities of electricity will be required for industrial use. Of particular importance to the question of emission control in the metals industries, however, is the selection of technologies which can both reduce the emissions of SOx and NOx whilst also leading to energy saving.

From data presented by Bushuev (1993), however, some 130-170TWh could be saved through Russian energy conservation programmes financed through financial incentives, credit options, government subsidies and tax options, providing that strict energy saving standards are put in place. The majority of savings would be expected from industry (85-105TWh), mainly from engineering (27-31TWh) and metallurgy (11-16TWh). The domestic sector would also provide savings of some 38-53TWh and the power generation sector itself would provide savings of some 3-4 TWh. Clearly, the content of the energy saving programmes would vary between industry, where attention would be focused on process control, and the domestic sector where attention would be paid to metering and temperature control of living accommodation.

The power generation sector does not appear to be a priority candidate for savings in electrical power itself, but it is an important industry for possible fuel savings. It is estimated that savings of 10-13 billion m^3 (bcm) of natural gas can be made in power generation, within a total national savings target of 30-35bcm, and 10 million tons of coal within a total national target of 28-34 million tons. These anticipated savings account for some 16-25 per cent of anticipated gas consumption by the power generation sector in the year 2000, and some 18-29 per cent for coal consumption. It is apparent, therefore, that scope exists for the transfer of Western technologies related to energy conservation, and improved power generation operation to obtain consequent fuel savings.

CONCLUSIONS

The information presented in this chapter has provided a means for selecting the possible combustion technologies which could be transferred to Russia and other republics of the former USSR, as a means of reduction of pollutant emissions during the fossil fuelled power generation process. Table 4.5 below is a summary of that data, indicating whether the technology has been proven, and also its level of expense. It is probably technically feasible to transfer all of these technologies to the former USSR, although there could be commercial problems related to intellectual property rights, but a selection has consequently been made on the factor that the technology has been proven.

Table 4.5 *Summary of Characteristics of Low Pollutant Emission Technologies*

AREA OF APPLICATION	FUELS		COMBUSTION				
Technology	Use of Lower Sulphur Fuels	Fuel Cleaning	Low NOx Burners	LIMB Process	Combined Cycles	Integrated Gasification Combined Cycle	Fluidised Beds
Proven Technology	Yes	Yes	Yes	No	Yes	Not widely proven	Not widely proven
Cost	Moderate	Moderate	Moderate	Unknown	Moderate	Not fully known	Not fully known

Table 4.5 continued

AREA OF APPLICATION	POST-COMBUSTION		IMPROVED EFFICIENCY	ENERGY SAVING
Technology	FGD	SCR	Operating Procedures	Operating Procedures
Proven Technology	Yes	Yes	Yes	Yes
Cost	High	High	Low	Low

Using these parameters as a basis for selection, therefore, it is apparent that the most obvious candidate for technology transfer is combined cycle technology in view of this technology's high efficiency and cleanliness, and the scope for using available gas supplies in Russia in a more effective fashion. It is apparent that the inlet temperature parameters of turbines designed in the former USSR have been lower than their counterparts produced in advanced Western companies, and hence there is scope for the transfer of Western technologies related to gas turbines and combustion chamber design (see Chapter 6). Other strong candidates for technology transfer include fuel cleaning, low NOx burners and improved efficiency, in view of the effectiveness and comparative inexpensiveness of these methods, and their scope for retrofitting to existing plants subject to coal combustion properties (particularly in Siberia) being suitable for these techniques.

AFBC, PFBC, IGCC and LIMB systems could also be considered for future development, dependent upon the levels of expertise in the former USSR in these fields of technology. FGD and SCR have been found to be extremely effective from the technical viewpoint, but they are also very expensive to install and operate; although FGD technologies may be required in some regions of the former USSR since some supplies of coal have a high sulphur content, and FGD and SCR systems may be installed more cheaply in the former USSR than in the West because of differences in labour costs. The introduction of clean coal technologies could be of particular importance in Ukraine, in view of the limited levels of oil production and gas reserves, although investment in improved mining techniques will be required to exploit available coal resources in that republic.

In view of the current political and economic conditions in the former USSR, therefore, and the limited funds available for import of technology, it is apparent that an immediately effective method for reducing atmospheric emissions of NOx and SOx is through energy saving at large factories, together with care taken in power station operation to reduce the levels of inputs and emission outputs during power generation. Furthermore, the levels of investment in equipment are comparatively small, as the majority of savings can be achieved through improved procedures and management.

In the medium and longer term, it will be necessary to assess future demands for electric power, due the recent fall in industrial output. This demand can then be compared with available capacity (see Chapter 5), with judgements made as to which power stations are likely to continue production, bearing in mind their size, age, location and type of fuel used and the probable necessity of maintaining combined heat and power stations in order that sufficient heat is provided to the population at large. Once a programme has been established with regard to possible closures and justifiable power station investment, it should then be possible to consider the various technological options for reduced emissions and the investments required. These capacity and investment-related issues will be discussed in more detail in Chapter 7.

NOTES

1. Boehmer-Christiansen and Skea (1991, p. 45). Other comparative cost data (1986 prices) exists for a proposed FGD plant for 2,000MW$_e$ power station of £30.22M for base equipment. £102.6M total erected cost (£51.29/kW$_e$), capital cost of £136M for new plant and £202M for retrofit, and operating cost of £24.02/kW$_e$.
2. Other estimates are provided by DoE (1991) which quotes an annual cost figure of £0.41/kW$_t$ for low Nox burners in new boilers compared with £18-53 per kWt for selective catalytic reduction. The costs per unit of emission abated was some £14/ton for low NOx burners in new boilers (45-50 per cent reduction) and £557-£1900 per ton (80-90 per cent reduction) for selective catalytic reduction.

REFERENCES

ABB promotion sheet, *PFBC: Clean Coal Technology*.

ABB, *ABB-EV Burner* (Trade Catalogue).

Akimov, N.K., Berezinets, P.A., Vasilev, M.K., Alekso, A.I., Gvozdev, A.V., Ershov, Yu.A., Ol'khovskii, D.D., Petrov, Yu.V., Proskuryakov, G.V., Barinberg, G.D. and Chukanov, A.D. (1992), 'Teplofikatsionnaya parogazovaya ustanovka moshnostyu 130MVt', *Teploenergetika*, No. 9, pp. 22-27.

Allen, J. (1990), 'Low NOx Burner Systems', *Energy World*, November, pp. 13-5.

Aminov, R.Z., Koval'chuk, A.B., Doronin, M.S., Shcheglov, A.G., Borisenkov, A.E., Zabuga, A.A. and Shaufler, L.G. (1994), 'O konversii moshchnykh aviatsionnykh gazoturbinnykh dvigatelei dlya statsionarnoi energetiki', *Teploenergetika*, No. 6, pp. 59-62.

Batenin, V.M., Zeigarnik, Yu.A., Kopelev, S.Z., Maslennikov, V.M., Novikov, A.S., Polezhaev, U.V., Favorskii, O.N. and Shternberg, V.Ya. (1993), 'Parogazovaya ustanovka s vvodom para v gazovuyu turbinu - perspektivnoe napravlenie razvitiya energeticheskikh ustanovok', *Teploenergetika*, No. 10, pp. 46-52.

Boehmer-Christiansen, S. and Skea, J. (1991), *Acid Politics: Environmental and Energy Policies in Britain and Germany*, London: Belhaven.

Bushuev, V.V. (1993), 'Novaya energeticheskaya politika Rossii (osnovnye polozheniya kontseptsii)', *Energetik*, No. 5, pp. 4-9.

Carette, Y.M. and McMillan, J. (1990), 'Environmental aspects of gas turbine combined cycle', *Power Generation and the Environment*, I.Mech.E./M.E.P., pp. 135-144.

Dawes, S.G. (1990), 'PFBC development - an overview of the achievements', *Energy World*, November, pp. 9-12.

Dawes, S.G., Brown, D. and Hyde, J.A.C. (1990), 'Options for advanced power generation from Coal', *Power Generation and the Environment*, I.Mech.E/M.E.P., pp. 123-34.

Department of the Environment (DoE) (1991), *Manual of Acidic Emission Abatement Technologies, Volume 1: Coal-Fired Systems*, London: HMSO.

Dibelius, G. (1988), 'Combined Cycle Plants', in VDI Berichte, *Combustion Pollution Reduction: New Techniques in Europe*, Dusseldorf: VDI Verlag, pp. 259-77.

Fillipov, G.A. (1990), 'Ekologicheskie aspekty v energetike i mashinostroenii', *Tyazheloe mashinostroenie*, No. 9, pp. 2-6.

Gribkov, A.M., Galiev, I.G., Chadaev, A.V., Uvarov, M.N., Zamal'dinov, Kh. A., Zolotov, A.P. and Shabarov, Yu.F. (1993), 'Snizhenie vybrosov oksidov azota na kotle TGM-84B', *Teploenergetika*, No. 9, pp. 40-3.

Haupt, G., Zimmermann, G., Meyer, B. and Schulze, O. (1995), 'Can hot gas clean up improve IGCC efficiency today?', *Power-gen Europe 95 Conference Papers, Volume 4*, Amsterdam, pp. 49-75.

Heppenstall, T. and Trimm, D. (1990), 'Practical options in power generation and their environmental impact', *Power Generation and the Environment*, I.Mech.E./M.E.P., pp. 77-84.

International Energy Agency (IEA) (1988), *Emission Controls in Electricity Generation and Industry*, Paris: OECD.

International Energy Agency, (IEA) (1994), *Electricity in European Economies in Transition*, Paris: OECD.

International Energy Agency (IEA)/Organisation for Economic Co-operation and Development (OECD) (1993), *Clean Coal Technology: Options for the Future*, DTI/ETSU.

Joyce, J.S. (1993), 'How Gas Turbines Can Improve the Operating Economy and Environmental Compatibility of New and Old Steam Generating Stations', Siemens AG Power Generation Group, 27pp. (reprinted from *Siemens Power Journal*, December 1992).

Kotler, V.R. and Enyakin, Yu.P. (1994), 'Realizatsiya i effektivnost tekhnologichesiikh metodov podavleniya oksidov azota na TES', *Teploenergetika*, No. 6, pp. 2-9.

National Power Environmental Performance Review 1991.

Newbery, D.M. (1993), 'The Impact of EC Environmental Policy on British Coal'. *Oxford Review of Economic Policy*, Vol. 9, No. 4, pp. 66-95.

Okhotin, V.N. (1992), 'Problemy okhrany okryzhayushchie sredy po proektirovanii TES', *Energetik*, No. 6, pp. 4-6.

Orlov, V.N. (1992), 'Gazoturbinnyi dvigatel' aviatsonnogo tipa NK-37 dlya elektrostantsii', *Teploenergetika*, No. 9, pp. 27-31.

Osyka, A.S. and Efimov, V.S. (1994), 'K voprosu o modernizatsii energeticheskoi gazovoi turbinoi GT100 po LMZ', *Elektricheskie stantsii*, No. 3, pp. 11-16.

Oudhuis, A.B.J., Yansen, D., Iwanski, Z. and Golec, T.W. (1993), 'Repowering of Polish coal-fired plants with IGCC', Paper presented at the Twelfth EPRI Conference on Gasification Power Plants, 27-29 October 1993, Hyatt Regency, San Francisco, USA.

'Perspektivy primeneniya gazovykh turbin v energetike', *Teploenergetika*, 1992, No. 9, pp. 2-8 (no author cited).

Pitrolo, A.A. and Bechtel, T.F. (1988), 'Simplified IGCC: Coal's "Adam Smith" Response to a Changing World', *Proceedings of the Seventh Annual EPRI Contractors' Conference on Coal Gasification*, EPRI AP-6007-SR, October, pp. 51-8.

Polishchuk, V.L. (1993), 'Voprosy razvitiya energeticheskogo gazoturbino-stroeniya i sozdaniya perspektivnykh gazoturbinnykh sistem novogo pokoleniya', *Teploenergetika*, No. 12, pp. 42-8.

Prutkovskii, E.N., Safonov, L.P. and Tarasov, E.A. (1990), 'Perspektivnye parogazovye ustanovki dlya elektrostantsii', *Tyazheloe mashinostroenie*, No. 5, pp. 8-12.

Prutkovskii, E.N. and Chavchanidze, E.K. (1995), 'Combined Cycle Steam and Gas Units with the Clean-up of the Flue Gases from Carbon Dioxide', *Heat Recovery Systems and CHP*, Vol. 15, No. 2, pp. 215-30.

Raymant, A.P. (1990), 'NOx reduction from new and existing power plant - a review', *Power Generation and the Environment*, I.Mech.E/M.E.P., pp. 37-41.

Redman, R. (1989), 'Fluidised bed combustion, SOx and NOx', *The Chemical Engineer*, December, pp. 32-8.

Rodgers, L.U., Morris, T.A. (1994), 'Snizhenie vybrosov oksidov azota topochnymi metodami', *Teploenergetika*, No. 6, pp. 10-15.

Rolls Royce International Power Group International Combustion (RRIPG), *Low NOx Combustion* and Babcock Energy *Low NOx Axial Swirl Burner*.

Romanov, V.I., Rudometov, S.V., Zhiritskii, O.G. and Romanov, V.V. (1992) 'Novyi gazoturbinnyi dvigatel' moshchnost'yu 110 MVt dlya statsionarnykh energeticheskikh ustanovok', *Teploenergetika* No. 9, pp. 15-21.

Schemenau, W. and van den Berg, C. (1990), 'The future of coal-fired power plants', *Power Generation and the Environment*, I.Mech.E./M.E.P., pp. 1-12.

Shirokov, S.N. (1990), 'Sostoyanie i perspektivy upravlenie dioksida sery iz dymovykh gazov teplovykh elektrostantsii', *Tyazheloe mashinostroenie*, No. 9, pp. 13-15.

Siemens Power Generation (1994), *Gas Turbines and Gas Turbine Power Plants*.

Sitas, V.I. and Streker, E. (1994), 'Opyt raboty konsaltingovykh firm v oblasti energosberezheniya i ekologii na primere firmy LINDeN (Velikobritaniya)', *Teploenergetika*, No. 3, pp. 70-3.

Sokolov, E.Ya. (1993), 'Razvitie teplofikatsii v Rossii', *Teploenergetika*, No. 12, pp. 2-7.

Sondreal, E.A. (1992), 'Clean Utilization of Low-Rank Coals for Low-Cost Power Generation', Proceedings of the Energy and Environment: Transitions in Eastern Europe Conference, Volume 1 (Sessions A1-5), Prague, 20-23 April, pp. 53-87.

Sondreal, E., Jones, M., Hurley, J., Benson, S. and Willson, W. (1994), 'Impact of Fuel Properties on Advanced Power Systems', Paper presented at the Second International Conference on Energy and Environment: Transitions in East Central Europe, Focus on Technology and Socioeconomic Issues, Prague, 1-5 November.

Veshnyakov, E.K., Varfomoleev, Yu.I. and Dmitreva, R.L. (1990), 'Vybor sposobov ochistki dymovykh gazov energeticheskikh kotlov ot oksidov sery i azota', *Tyazheloe mashinostroenie*, No. 9, pp. 1, 5-8.

Zarubin, L.A., Simma, F.Ya., Gorbachinskii, S.I., Shiliu, Yu.P. and Kolomiets, A.M. (1992), 'Parogazovaya ustanovka PGU 350 NPO Turboatom', *Teploenergetika*, No. 9, pp. 9-14.

5. Power Generation in the Former USSR

INTRODUCTION

Chapter 2 has provided information on the levels of atmospheric emissions in the former USSR, paying particular attention to estimates of SOx and NOx emissions from the power generation industry, followed by Chapters 3 and 4 which provided information on the properties of fuels and combustion processes used in that region and also in the West. This chapter provides a context for the environmental and technical information presented in the previous chapters by describing the organisational framework of the power generation industry, and analysing data on electricity output, levels of capacity and fuel mix within the industry. In addition, the information contained in this chapter also provides background information for the case studies of technology transfer in Chapter 6, and commercial information provided in Chapter 7.

THE SOVIET ORGANIZATIONAL LEGACY
(Whitefield, 1993; pp. 92-201; Cooper, 1991, pp. 6-11; Rosengaus, 1986; Rosengaus, 1987, pp. 1-51)

The major organisations concerned with power station pollutant emission in the former USSR can be divided into four main groups, namely power generators, equipment-providers, fuel providers, and public health and environmental organisations. Each of these four main groups of organisations were previously controlled by ministries or state committees having administrative authority throughout all of the constituent republics of the former Soviet federal Union.

Power generation from thermal power stations burning fossil fuels and other hydrocarbons has accounted for the majority of electrical production in the former USSR (some 70-75 per cent), with hydro-power accounting for some 15-17 per cent of output. These power stations were the responsibility of the Ministry of Energy (*Minenergo*), which co-ordinated production through local, area and republican organisations; and distribution was carried out on local and regional bases, with interconnections into a federal network. *Minenergo* also had a number of research and development organizations responsible for technological investigations in power generation and distribution. The generation of nuclear power (some 10-12 per cent of electrical output) was the responsibility of the Ministry of the Atomic Energy Industry (*Minatomenergoprom*) which was created in 1989 from a merger of the Ministry for Nuclear Power previously responsible for civilian nuclear power generation; and the Ministry for Medium Engineering, which was previously responsible for the Soviet nuclear weapons programme. Public health and environmental organizations responsible for monitoring environmental pollution, including atmospheric emissions from power stations, were either responsible to the Ministry of Health, local councils or the State Committee for the Environment (*Goskompriroda*).

The providers of the fuel required to generate electrical power were divided into four main groups based on the type of fuel, namely coal mines responsible to the Ministry of the Coal Industry (*Minugleprom*), gas producers responsible to the Ministry of the Gas Industry (*Mingazprom*), oil producers responsible to the Ministry of the Petroleum Industry, (*Minneftprom*), and nuclear fuel producers responsible to the Ministry of the Nuclear Energy Industry (*Minatomenergoprom*).

The major manufacturers of the power generation equipment required to convert thermal energy to electrical energy were responsible to the Ministry of the Power Engineering Industry (*Minenergomash*), including most of the factories and research and design establishments engaged in boiler and turbine technology for the power generation industry; although some other manufacturers of power plant were also responsible to the Ministry of Heavy Engineering (*Mintyazhmash*). Factories responsible to these ministries would carry out power engineering production for the nuclear power industry, but a substantial proportion of these items would have been made in facilities responsible to the Ministry of the Nuclear Energy Industry itself. Electrical generators and distribution equipment were produced by factories responsible to the Ministry of Electrical Engineering (*Minelektrotekhniki*), whilst other ancillary equipment such as fuel and water treatment facilities would have been produced by factories responsible to the Ministry of Chemical Engineering (*Minkhimmash*), as well as factories responsible to the Ministry of Power Engineering.

In this chapter, attention is to be focused on the major participants in fossil fuelled electrical power generation, namely establishments responsible to the former Ministry of Energy (*Minenergo*). Particular attention will be paid in Chapter 6 to the research, development, design and production facilities previously responsible to the Ministry for Power Engineering (*Minenergomash*). It is also important to note, however, that significant development work in the various technical fields related to electric power generation has also been carried out at the Institute of Energy Problems (responsible to the Russian Academy of Sciences) and the Moscow Energy Institute (responsible to the Ministry of Higher Education).

POST-SOVIET ORGANISATION OF POWER GENERATION

As explained in the previous section of this chapter, the Ministry of Energy was formerly responsible for the generation and distribution of power on an all-Union basis, with the output of this homogeneous commodity defined by the parameters of frequency (cycles per second) and production (kilowatt hours [kWh] or terawatt hours [TWh]), generated from power stations having specified capacities in megawatts (MW). The majority of electrical power was generated and distributed on a regional or local basis, with widespread use of combined heat and power (or co-generation) facilities which accounted for approximately 20 per cent of the installed power generation capacity (Rosengaus, 1987, pp. 28-41) and some 34 per cent of the electricity produced during the 1970s. The maximum capacity of a co-generation plant was some 250MW (Hewett, 1984, pp. 114-16). Following the fragmentation of the former USSR into independent sovereign states in 1991, power generation facilities on Russian territory were administered by the newly created Ministry of Fuels and Energy (*Mintopenergo*) which has retained responsibility for strategic development of the fuel and energy industries (*Elektricheskii stantsii*, 1993).

Electrical power in Russia at the beginning of 1993 was generated from 728 power stations, of which 618 were thermal power stations, with the remainder being hydroelectric stations (101) and nuclear power stations (9). These installations were networked into 72 power distribution systems, 66 of which were connected into 7 linked systems, of which 6 worked in parallel and formed the basis of a Unified Electrical System. The total capacity of Russian *Mintopenergo* power stations was some 173,000MW in 1993, comprising some 130,000MW (75 per cent) installed in thermal power stations and 43,000MW (25 per cent) in hydro-electric power stations. Atomic power stations operating in the Unified Electrical System accounted for an additional capacity of some 20,000MW. These capacity levels are almost identical to those of 1990 for hydro and nuclear power, but capacity levels in thermal power stations had fallen from a 1990 level of 148,400MW, according to data presented in a fuel and energy map compiled by an authoritative research institute responsible to the Russian Ministry of Economics.[1] The production from Russian electricity generating capacity in 1992 was some 1009TWh in total, including some 717TWh from thermal power stations, 173TWh from hydro-electric stations, and 119TWh from nuclear power stations (Red'ko, 1993).

The process of privatization removed 51 large state regional power stations (*gosudarstvennye regional'nye* (or *raionnye*) *elektrostantsii* [*GRES*]) from within 31 of the 72 former regional power systems. These 51 large power stations have a capacity of 1,000MW and greater for thermal power stations, or 300MW and greater for hydro-electric stations, and form the hubs of subordinate organizations of the state-owned Russian shareholding companies for electrical energy, linked together through the national grid, usually referred to as the 'Unified Electrical System'. This capacity accounts for some 96,000MW in total (suggesting an average power station capacity of some 1,900MW), although another source refers to some 50,000MW for the total capacity of large power stations supplying the national grid (IEA, 1994, p. 191). This suggests that some 77,000MW would remain in the 72 regional distribution systems, if the total capacity of the large power stations feeding the national grid is taken to be some 96,000MW, or some 123,000MW if the total capacity of large power stations is taken to be some 50,000MW. These regional distribution systems have been converted into regional joint stock companies with the right to produce electricity from their own power stations or to purchase this commodity in the national market. Of these 72 regional systems, 20 were expected to be 'energy-rich' (that is, having surplus generating capacity in the region), and the remainder were expected to be 'energy-deficient', with the latter having the right to share the profits obtained by purchasing from local power stations at tariffs lower than those of the national grid (Red'ko, 1993; Baranovskii, 1993; D'yakov, 1993). A state-owned joint stock company (*RAO-'EES Rossii'*, or Russian Joint Stock Company-'Unified Electrical System of Russia') co-ordinates and implements state policy on national electrical energy strategy. This has been achieved through the purchase of 49 per cent of the foundation capital of each of the regional joint stock companies for electrical energy and assimilation of the large power stations into the state grid (Zagyanskii and Red'ko, 1994).

An example of this process is provided by the reorganization of the *Rostovenergo* system in the Rostov-on-Don region of Southern Russia. *Rostovenergo* was the fifteenth largest power system in the former USSR with a capacity of more than 3,000MW from 6 electrical power stations and 2 combined heat and power plants stations, serving a population of 4 million people resident in an area of 100,000km^2. 2245MW of this capacity was previously supplied by the Novocherkass State Regional Power Station *(GRES)*, with a further 200MW from the Tsimilanskaya hydro-electric station and several combined heat and power stations. The 300 and 500KV lines and the Novocherkass State Regional Power Station were assimilated into RAO-EES Rossii, and the remaining capacity of *Rostovenergo* was available for lease. *AO Rostovenergo* was then established for regional electricity distribution and to manage this remaining capacity, with leasing arrangements for capacity in the Novocherkass Power Station and the use of a small combined heat and power plant (6MW) at the Rostsel'mash agricultural engineering plant. 49 per cent of the shares of *Rostovenergo* are retained by *RAO EES* for three years, and the remaining 51 per cent are distributed between the labour force (35 per cent), the property fund of the Rostov region (10 per cent), and purchases made at auctions (6 per cent) (Kushnarev, 1994).

From this survey of the process of privatisation of electric power generation utilities in Russia therefore, it is apparent that the key decision-makers in the selection and implementation of pollution-reduction equipment in power stations will be the individual power stations, and the directorates of the local 'energy systems' (or distribution companies) in which power stations are located. The power generation directorates of *RAO-EES Rossii* as the significant shareholder in all of the local energy systems and as the owners of the large thermal and hydro power stations providing electricity into the national grid, will also play important roles in the selection of pollution control equipment and could also exert an additional influence through the development of a policy of preferred suppliers.

In addition to its production capability, the former Ministry of Fuels and Energy also co-ordinated a number of research and development establishments, including the internationally famous All-Union Thermal Energy Institute *(Vsesoyuznyi teploenergeticheskii institut (VTI))*. This institute has now been re-named as the All-Russian Thermal Energy Institute, and has presumably been privatised in common with other research and development establishments in the Russian civilian sector. In view of this institute's many years of experience in the selection of various technologies for power generation, it will probably continue to exert a major influence on the choice of Western power engineering technologies for the reduction of atmospheric pollution.

The progress towards privatisation has been slower in the other European former Soviet republics than in Russia, however. Belarus, Estonia, Latvia, Lithuania, Moldova and Ukraine have all retained a state monopoly of power generation administered through an industrial ministry reporting to a senior government body. Ukraine envisages the privatisation of some 4-6 joint stock companies for power generation, although the implementation of this plan is contingent upon prolonged stabilistation in the Ukrainian economy (IEA, 1994, pp. 239-55). The combined capacity remaining within a state monopoly for these European former Soviet countries is some 70,000MW, with Ukraine accounting for some 53,000MW of this capacity.[2]

POWER STATION CAPACITIES, AGES, AND FUEL MIX

Moscow Energy Institute (MEI) Data for Former USSR

Information on capacities and fuel mix in the power generation industry in the former USSR is available from two lists supplied by senior personnel from the Moscow Energy Institute in 1994 as part of a consultancy project for a UK power engineering company. This information therefore probably relates to 1992 or 1993. The first of these lists provides data on thermal power station location, date of installation, turbine model, type of fuel, average load, specific fuel consumption, and type of boiler and efficiency, for 74 installations of power generation sets between 150MW and 1200MW. The second list provides capacity data on turbines installed in 235 thermal power stations. The total capacity for these power stations can therefore be calculated from the data on capacity of each turbine and the quantity of turbines in each power station. Of the 235 power stations in the second list, 75 were state regional power stations connected to the Soviet national grid, whilst 160 were combined heat and power stations. Of the 75 state regional power stations included in the second list, 63 were shown in the first list, together with 6 combined heat and power stations. The difference between the 69 named power stations in both lists, and the 74 installations in the first list, is accounted for by double counting of 8 power stations which were mentioned at least twice in the first list as that list categorised power stations by capacity of turbine set (between 150MW and 1200MW) rather than total power station capacity. In addition, three power stations were mentioned in the first list, but not in the second.

The data available from these two lists were merged to provide information on total capacity and fuel mix capacity for a total of 72 thermal power stations ranging from 150MW to 4,800MW. The total capacity of these 72 power stations was some 117,000MW, which accounted for approximately 50 per cent of the total electricity generation capacity from thermal power stations in the former USSR in 1990 (240,800MW according to the *Toplivno Energeticheskii Kompleks*). This sample accounted for less than 10 per cent of the total quantity of power stations in the former USSR, however, and it has been assumed in this chapter that they are therefore the largest 10 per cent of thermal power stations in that region. These data have been grouped into four capacity categories, namely 150-999MW, 1000-1999MW, and 2,000-2999MW, and 3,000MW and greater, and classified by fuel mix and age as shown in Table 5.1A appended to this chapter. The capacity for each type of fuel within each of the four capacity categories has been calculated from the fuel percentage given for each power station in Table 5.1A and then aggregated within each total turbine capacity category. The results of these calculations are shown in Table 5.1 for the sample of power stations which ranges from 150MW to 4,800MW.

Table 5.1 Fuel Mix and Power Station Capacity (MW)

Power Station Capacity Category	150-999MW	1000-1999MW	2000-2999MW	3000 MW+	Total
Gas-fired Capacity (MW)	6,704	16,406	11,097	15,038	49,245 (42%)
Coal-fired Capacity (MW)	1,913	18,174	12,554	11,448	44,089 (38%)
Oil-fired Capacity (MW)	1,223	7,610	8,965	6,378	24,176 (20%)
Total Capacity (MW)	9,840 (8%)	42,190 (36%)	32,616 28%)	32,864 (28%)	117,510 (100%)

From Table 5.1 it can be seen that the distribution of total turbine capacities was similar across the three groupings from 1,000MW to 3,000MW and greater (36 per cent, 28 per cent and 28 per cent respectively), but that the 150-999MW capacity group only accounted for 8 per cent of the total. From this table, it can also be seen that for this sample of power stations of 150MW and greater, 42 per cent (that is, 49,245MW) of the total capacity is gas-fired, 38 per cent (that is, 44,089MW) is coal-fired, and 20 per cent (or 23,176MW) is oil-fired.

These aggregate figures hide large differences in fuel combustion, however, between fuels used in different sizes of power station, and in power stations of different age groupings. Gas, for example, is used to fuel 68 per cent (6704/9840) of the capacity in power stations of 150-999MW, reducing to 46 per cent (15038/32864) of the capacity in power stations of 3,000MW and greater; and some 37 per cent (16406 + 11097)/(42190 + 32616) for power stations of capacity between 1000MW and 2,999MW. Coal, on the other hand accounts for only some 20 per cent of capacity in power stations of 150-999MW, but 35 per cent of capacity in power stations of 3000MW and greater, and 41 per cent of capacity in power stations within the capacity range of 1,000MW-2,999MW. From these data, therefore, it appears that gas is by far the most favoured fuel in power stations of 150-999MW capacity and 3000MW and greater, whereas coal is marginally the most favoured fuel (closely followed by gas) in power stations of 1,000MW-2,999MW capacity. Although oil is used as the major fuel in only two of the nine largest power stations, it appears to be widely used as a secondary fuel in power stations over a range of capacities. Furthermore oil is used to fuel only 12 per cent of capacity for power stations of 150MW-999MW capacity, although it is used to fuel some 27 per cent of the capacity for power stations between 2,000 and 2,999MW.

A further factor to be considered is the age pattern of power stations. For very large power stations (3,000MW and greater), some 97 per cent of the gas-fired capacity is less than 20 years old, with some 32 per cent less than 10 years old; and some 94 per cent of oil-fired power station capacity, and 81 per cent of coal-fired power station capacity are less than 20 years old (see Table 5.2). For power stations of 150-999MW capacity, however, only 29 per cent of gas-fired capacity is less than 20 years old, and only 22 per cent of oil-fired capacity is within this age category compared with 87 per cent of coal-fired capacity. For power stations between 1,000-3,000MW, only some 5 per cent of coal-burning capacity and 10 per cent of oil burning capacity are less than 10 years old, compared to some 33 per cent of gas-burning capacity. From these data, therefore, it appears that very large power stations (3,000MW +) contain the most modern capacity compared with other sizes of power stations, whilst gas-fired power stations appear to be the most modern types between 1000 and 3,000MW + capacity but coal-fired stations appear to be the most modern types for power utilities of 150-999MW. Power generating capacity in power stations of 1,000-3,000MW appears to be the most old-fashioned, especially for coal and oil-fired stations.

From the information shown in Table 5.1A it is also possible to highlight existing coal-fired power stations which may be selected for the installation of atmospheric pollution control equipment. Pollution control equipment for the reduction of SOx emissions could be fitted to the Ekibastuz and Reftinsk power stations, for example, as these have high capacities over which the high investment costs could be absorbed, and the age of these power stations (13 years and 17 years respectively) is such that the additional investment should have a sufficiently long working life. Combustion data for Ekibastuz coals indicates very high levels of SOx emissions (see Chapter 3), and these fuels are probably burnt in both the Ekisbastuz and Reftinsk power stations. Data available from Table 5.1A also suggests that the Ryazanskaya coal-fired power station is worthy of further investigation in view of its large capacity (2,800MW of which 79 per cent is coal-fired), although the station is some twenty two years old. The Gusinoozersk power station is only some 13 years old, however, has a capacity of some 1050MW and is located relatively near to the large water resource of Lake Baikal.

Such SOx and NOx reduction projects would probably require some regionally-based cross-subsidy to pay for the imported equipment and technology. Some efficiency gains should be obtained from the use of more modern equipment (for example, low NOx burners), but market development would be necessary to sell the coal thereby saved, or the savings in industrial materials received as payments in kind from industrial customers, to Western countries. Some types of pollution control equipment (for example, FGD and SCR) can be very expensive, however, with no perceived savings in energy efficiency unless attention is paid to improved power station operation; and the installation of such equipment would therefore require external support from Western unilateral or multilateral aid. Many of the remaining larger coal burning power stations, are probably too old for the installation of pollution control equipment to be economic, although there may be some scope for the fitting of low NOx burners in the smaller and more modern coal-fired power stations. Programmes may also be developed for the installation of low NOx burners and FGD and SCR

Table 5.2 Fuel Mix, Power Station Capacity and Age of Capacity

Power Station Capacity Category		0-999 MW			1000-1999 MW		
Fuel	Age of Power Station (Yrs)	No. of stations	Capacity (MW)		No. of stations	Capacity (MW)	
	0-10	1	590	(9%)	3	3,136	(19%)
	11-20	2	1,330	(20%)	2	2,885	(18%)
Gas	21-30	2	774	(12%)	13	9,869	(60%)
	31-40	8	4,010	(60%)	1	516	(3%)
	Sub-total	13	6,704	(100%)	19	16,406	(100%)
	0-10	1	720	(37%)	2	792	(4%)
	11-20	2	972	(50%)	1	893	(5%)
Coal	21-30	0	0	(0%)	9	7,848	(43%)
	31-40	3	221	(11%)	7	8,641	(48%)
	Sub-total	6	1,913	(100%)	19	18,174	(100%)
	0-10	1	80	(7%)	2	72	(1%)
	11-20	4	188	(15%)	1	157	(2%)
Oil	21-30	2	406	(33%)	16	6,737	(88%)
	31-40	5	549	(45%)	4	644	(9%)
	Sub-total	12	1,223	(100%)	23	7,610	(100%)

Table 5.2 continued

2000-2999 MW			3,000 MW+			Total Capacity (MW)	
No. of stations	Capacity (MW)		No. of stations	Capacity (MW)			
1	1,218	(11%)	1	4,800	(32%)	9,744	(20%)
1	1,752	(16%)	4	9,788	(65%)	15,755	(32%)
5	7,327	(66%)	1	450	(3%)	18,420	(37%)
2	800	(7%)	-	-		5,326	(11%)
9	11,097	(100%)		15,038	(100%)	49,245	(100%)
-	-		6	-		1,512	(4%)
-	-			9,258	(81%)	11,123	(25%)
5	10,434	(83%)		2,190	(19%)	20,472	(46%)
2	2,120	(17%)		-		10,982	(25%)
7	12,554	(100%)	4	11,448	(100%)	44,089	(100%)
1	882	(10%)	-	-		1,034	(5%)
1	648	(7%)	6	6,018	(94%)	7,011	(30%)
9	6,355	(71%)	1	360	(6%)	13,858	(57%)
2	1,080	(12%)	-	-		2,273	(9%)
13	8,965	(100%)	7	6,378	(100%)	24,176	(100%)

equipment in new coal-fired power stations, with some new projects being directed towards the use of advanced coal burning technologies such as circulating circulating fluidised beds (CFB) and integrated gasification combined cycles (IGCC). Fluidised bed combustion technology may be investigated for installation at some of the smaller and newer coal-fired power stations, such as the 390MW *Neryungrinskaya* utility, constructed in 1983, particularly as Neryungren coals burn with low levels of SOx (see Chapter 3).

In the case of gas-fired power stations, a wide range of power stations of various sizes and of service life of less than 20 years are available for possible renovation. The selection of locations is therefore likely to be influenced by commercial considerations such as potential for sale to the West of electricity generated at the refurbished station, or the potential for sale of other commodities from that region in Western markets. These sales might include some of the savings in natural gas from the use of higher efficiency equipment, or industrial materials and other commodities received as payment in kind from some major customers. It is likely that refurbishment of gas-fired power stations will consist of either the fitting of low NOx burners to existing boilers, which is comparatively inexpensive and cost-effective for recently-installed boilers and engineers in the former USSR have established expertise in the development and installation of these items (see Chapter 4); or the replacement of the gas-fired boiler by a gas-turbine and heat recovery steam generator for use in a combined cycle configuration.

In view of Russian designs of large power generation gas turbines being less technically advanced than those of their Western counterparts, however (see Chapter 4), there is potential for the transfer of Western technology for these types of equipment, which is being exploited by both Western and Russian companies (see Chapter 6). In addition, there is potential for the use of aero-derivative turbines of up to 40-50MW capacity in some of the smaller power stations, providing that the installations are not too remote to prevent adequate maintenance and servicing from being achieved. Western companies have established expertise in these installations, and factories in Rybinsk (Russia) and Nikolaev (Ukraine) also have experience in the development of 12-16MW aero-derivative gas-pipeline pumping turbines, which may be transferred to power generation applications (see Chapter 4). Renovation of gas-fired power stations will probably be influenced more by economic than environmental factors, however (see Chapter 7), even though natural gas has inherent cleanliness as a fuel for power generation.

When using this information for evaluation purposes, however, it is important to bear in mind that these data come from one Russian source only, although that source is authoritative. When comparing these data with information provided in the energy map (*Toplivno-Energeticheskii Kompleks*) for the former USSR compiled in 1993 by the All-Russian Institute of Integrated Fuel and Energy Problems of the Russian Ministry of Economics, it is apparent, that there are major consistencies and only minor discrepancies between these two sources, for power stations of 2,000-2,999MW capacity (see Table 5.2A appended to this chapter). There are also close consistencies in the data presented for the 3,000MW+ capacity, apart from the data for the Uglegorskaya power station in the Donets province of Ukraine, where the Russian energy map provides a total capacity of 5,100MW for 3 power stations, compared with 3600MW cited by MEI. For the total capacity of 32,864MW presented in Table 5.2A for power stations of greater than 3,000MW capacity, there is a discrepancy of only some 1,656MW with the data presented in the Russian energy map. Data is also available from IEA (1994, pp. 212-23) for power stations in the European regions of the former USSR, classified by capacity and major fuels (see below); and the aggregate capacity information available for power stations of more than 3,000MW capacity provides only a difference of some 90MW compared to the MEI data within a total capacity of 25,864MW. This pattern is also repeated for power stations between 2,000 and 2,999MW where there is only a discrepancy of some 486MW between the MEI and the Russian energy map data in a total of 32,616MW, and only a discrepancy of some 329MW within a total of 28,116MW between the MEI and IEA data for the European regions of the former USSR.

For power stations of less than 2,000MW capacity, however, there are some wide variations between a few of the rated capacities quoted by MEI and the energy map, although there are also some very close consistencies. The reasons for the differences are not entirely clear, but they are probably as a consequence of differences of definition between power generation 'sets' and 'power stations', or the grouping of several power stations into one unit in the energy map. Comparisons with the IEA data present particular difficulties as none of the quoted power stations in Table 5.2A are in European Russia, and the IEA and MEI data refer to different minimum sizes of power station (see below), and identification of individual power stations from the energy map is sometimes difficult in the densely urbanised areas of the European Russian region. Nevertheless in several cases where comparison is possible (for example, Sredneural'skaya GRES, Mosenergo GRES 5, Yuzhnoural'skaya GRES, and Yuzhnaya TETs) there are no major variations between the MEI and IEA data in terms of capacities, although data sometimes differ on fuel mix.

International Energy Agency (IEA) Data for European Russia
(IEA, 1994, pp. 35-49, 81-91,111-51)

Data available from IEA (1994, pp. 212-23) for the capacity of power stations of 10MW and greater in European Russia can also be compared with the similar data available for the former USSR previously cited from the Moscow Energy Institute, although more capacity groups for stations of lower than 1,000MW have been used to present the IEA data in view of the large number and range of capacities of reported power stations. Furthermore, a category of 100-249MW has been introduced for the IEA data for comparing the power stations of this capacity within the various regions of European Russia, but this category has also been divided into two additional sub-categories (100-149MW and 150-249MW) for comparision with the MEI data, in view of the minimum capacity of 150MW quoted in that source. These data reveal a similarity in proportion between the 1,000-1,999MW, 2,000-2,999MW and 3,000MW+ categories for those power stations within the 1,000MW+ capacity range, although the quantities in each category are lower for the European Russian sample because of the smaller geographical region being considered (see Table 5.3). There are, however, large discrepancies in the data for the quantity of power stations of less than 1,000MW capacity. These differences can be explained through the selection of the definition of capacity (minimum of 150MW sets in the case of the Moscow Energy Institute data and 10MW in the case of the IEA [1994] data), which would obviously cause higher quantities in the 100-249MW range of 150MW and lower, but could also give rise to a higher proportion of power stations in the 250-499MW range because of the potential for configuring power generation sets smaller than 150MW to achieve a total capacity of less than 500MW. The IEA data is extremely useful, however, for the provision of information on the large number (40 per cent) of power stations of less than 100MW capacity with almost 25 per cent having a capacity of less than 50MW (see Table 5.3).

For the other European republics in the former Soviet Union, the IEA data suggests that there are higher proportions in the larger capacity range than in the case of the Russian Federation. These differences are probably due to Belarus, Estonia, Latvia, Lithuania, Moldava and Ukraine being more geographically compact and the population less scattered than in the Russian Federation; with a large proportion of Ukrainian power stations being built on coalfields to exploit the local availabilities of supplies of this fuel. It might be the case, however, that the IEA data is incomplete for these republics, as reference is only made to 'principal power plants' in the IEA publication.

Table 5.3 Fuel Mix and Power Station Capacity (Comparison of Moscow Energy Institute and IEA Data)

Capacity of Power Station	Moscow Energy Institute data[1]		IEA data[2] (for Russian Federation)		IEA data for other republics in former European Soviet region[3]
3,000MW+	(17%)[4]	9 (12%)[5]	(11%)[4]	4 (1%)[5]	3 (6%)[5]
2,000-2,999MW	(35%)	14 (19%)	(31%)	11 (3%)	5 (11%)
1,000-1,999MW	(57%)	31 (42%)	(57%)	20 (5%)	10 (21%)
500-999MW		17 (23%)		38 (9%)	7 (15%)
250-499		1 (1%)		81 (19%)	6 (13%)
100-249		2 (3%)		98 (23%)	8 (17%)
[incl 150-249		2 (3%)		[34 (8%)	[7 (15%)
100-149]		0 (0%)]		64 (15%)]	1 (2%)]
50-99				68 (16%)	3 (6%)
0-49				99 24%)	5 (11%)
Total	(100%)	74 (100%)	(100%)	419 (100%)	47 (100%)

Notes:
1. See Table 5.1A for data provided by Moscow Energy Institute (MEI).
2. Aggregated from data presented in IEA (1994), pp. 212-23.
3. See IEA (1994), pp 35-188, 239-55.
4. That is 9/9 + 14 + 31 for MEI data, and 4/4 + 11 + 20 for IEA data.
5. That is 9/74 for MEI data, and 4/419 for IEA data etc.

The data provided in Table 5.4 is an aggregate for all of the five major European Russian regions and is dominated by data available for the Central region which accounts for some 151 power stations from the European Russian total of 419, and some 75,000MW capacity out of the European Russian total of 165,000MW (see Table 5.5). Central region is followed by the Urals region in terms of both number of power stations (98) and generating capacity (40,000MW), and both of these regions have a large proportion of their total quantity of power stations burning gas as their dominant fuel (60 per cent in the case of Central Region and 53 per cent in the case of Urals region), followed by coal (26 per cent in the case of Central region and 28 per cent in the case of Urals region) (see Tables 5.5 and 5.6). A large proportion of the power stations in the North Caucasus and Middle

Table 5.4 Fuel Mix, Power Station Capacity and Numbers of Power Stations for European Russia[1] (1993)

Type of Station	Hydro	Gas	Oil	Coal [2]	Total Number of Power Stations
Capacity Group					
3000 MW +	0	3	0	1	4
2000-2999MW	1	6	1	3	11
1000-1999MW	2	14	1	3	20
500-999MW	3	23	3	9	38
250-499MW	5	38	15	23	81
100-249MW	17	33	23	25	98
50-99MW	11	26	12	19	68
0-49MW	21	49	15	14	99
Total Number of Power Stations	60 (14%)	192 (46%)	70 (16%)	97 (23%)	419 (100%)

Notes:
1. Calculated from IEA data shown in IEA (1994) pp. 212-23.
2. Power stations burning peat (17 stations) and oil shale (2 stations) have been included in the 'coal' category, which has been used to include all solid fuels. 12 of the peat burning stations were in the Central Region, 2 in the North West, and 3 in the Urals. The distribution of the capacities for peat burning stations were: 250-499MW capacity, 2 stations; 100-249MW capacity, 7 stations; 50-89MW capacity, 3 stations; 0-49MW capacity 5 stations. For oil shale stations, there was one utility within the 100-249MW capacity, and another within the 0-49MW capacity.

Table 5.5 Fuel Mix, Power Station Capacity and Numbers of Power Stations for the Central Region of European Russia[1] (1993)

Type of Station	Hydro	Gas	Oil	Coal [2]	Total Number of Power Stations
Capacity Group					
3000MW +	0	1	0	0	1
2000-2999MW	1	3	0	0	3
1000-1999MW	0	9	1	3	13
500-999MW	0	20	0	3	23
250-499MW	2	20	1	12^2	35
100-240MW	1	14	4	10^3	29
50-99MW	1	10	4	8^4	23
0-49MW	2	14	3	4^5	23
Total Number of Power Stations	7 (5%)	91 (60%)	13 (9%)	40 (26%)	151 (100%)

Notes:
1. Calculated from data shown in IEA (1994), pp. 212-23.
2. Including 1 peat-fired station.
3. Including 5 peat-fired stations.
4. Including 2 peat-fired stations.
5. Including 4 peat-fired stations.

Table 5.6 Fuel Mix, Power Station Capacity and Numbers of Power Stations for the Urals Region of European Russia[1] (1993)

Type of Station	Hydro	Gas	Oil	Coal	Total Number of Power Stations
Capacity Group					
3000MW+	0	2	0	1	3
2000-2999MW	0	1	0	2	3
1000-1999MW	0	2	0	0	2
500-999MW	1	3	1	4	9
250-499MW	1	8	5	3	17
100-249MW	0	6	3	1[2]	20
50-99MW	0	8	1	4[3]	13
0-49MW	0	22	6	3	31
Total Number of Power Stations	2 (2%)	52 (53%)	16 (16%)	28 (29%)	98 (100%)

Notes:
1. Calculated from data shown in IEA (1994), pp. 212-23.
2. Including 1 peat-fired power station.
3. Including 2 peat-fired power stations.

Volga regions also consume gas (41 per cent and 52 per cent respectively, see Tables 5.7 and 5.8), but this fuel is marginally exceeded by hydro-generation (46 per cent of power stations) in the North Caucasus and followed by oil-firing rather than coal in the case of Middle Volga. The North West region contains the largest number of hydro power stations in the five European regions studied, accounting for almost one third of the power stations in that region, followed by oil (27 per cent of the total) and coal (22 per cent of the total) (see Table 5.9).

When an analysis of capacities is made, alongside a survey of numbers of power stations (see Table 5.10), it is apparent that only 8 per cent (or 35/419, see Table 5.4) of the total number of power stations have a capacity of greater than 1000MW but account for almost 50 per cent of power station capacity in European Russia (that is, 73,500MW/157,100MW, see Table 5.10) and gas is by far the most dominant fuel to be found in those stations particularly in the Urals region. On the other hand, some 40 per cent of the total power stations in European Russia have a capacity of less than 100MW (167/419, see Table 5.4) but only provide some 5 per cent (7,575/157,100, see Table 5.10) of the total generating capacity for that region; and less than 2 per cent of total generating capacity is provided by some 24 per cent of the total power stations which have a capacity of less than 50MW. It is also apparent from Table 5.10 that gas is the dominant fuel in terms of capacity (56 per cent of total) for European Russia, followed by coal (24 per cent of total). Hydro-power only accounts for some 9 per cent of the total capacity for European Russia, with oil accounting for some 11 per cent, although hydro-power accounts for some 21 per cent of the total capacity in power stations of less than 50MW.

The IEA data, therefore, serves as a useful complement to the MEI data previously referred to, as the MEI data highlights the capacity, age and fuel mix of power stations having turbine sets of greater than 150MW capacity, whereas although the IEA data only provides capacity and fuel mix data for European Russia alone, they give more detailed information on the capacity and fuel mixes of power stations of less than 150MW capacity. Furthermore, the IEA data provide capacity information on power stations which are located in areas of particular ecological concern to European countries (for example, the Petrozavodsk and Novgorod power stations near to the Finnish border).

From these data, it can be seen that as well as gas being the major fuel in aggregate in European Russian power stations, it is also the major fuel for power stations of less than 50MW capacity.

Table 5.7 Fuel Mix, Power Station Capacity and Numbers of Power Stations for the North Caucasus Region of European Russia[1] (1993)

Type of Station	Hydro	Gas	Oil	Coal	Total Number of Power Stations
Capacity Group					
3000MW+	0	0	0	0	0
2000-2999MW	0	1	0	1	2
1000-1999 MW	1	1	0	0	2
500-999 MW	0	0	0	0	0
250-499 MW	0	3	0	0	3
100-249 MW	4	0	1	0	5
50-99 MW	3	3	0	1	7
0-49 MW	6	5	1	1	13
Total Number of Power Stations	14 (44%)	13 (41%)	2 (6%)	3 (9%)	32 (100%)

Note:
1. Calculated from data shown in IEA (1994), pp. 212-23.

Table 5.8 Fuel Mix, Power Station Capacity and Numbers of Power Stations for the Middle Volga Region of European Russia[1] (1993)

Type of Station	Hydro	Gas	Oil	Coal	Total Number of Power Stations
Capacity Group					
3000MW	0	0	0	0	0
2000-2999 MW	0	1	0	0	1
1000-1999MW	1	2	0	0	3
500-999 MW	2	0	1	0	3
250-499MW	1	6	5	2	14
100-249 MW	0	7	6	0	13
50-99 MW	0	3	1	1	5
0-49 MW	0	5	1	1	7[2]
Total Number of Power Stations	4 (9%)	24 (52%)	14 (30%)	4 (9%)	46 (100%)

Notes:
1. Calculated from data shown in IEA (1994), pp. 212-23.
2. Including one oil-shale fired station.

Table 5.9 Fuel Mix, Power Station Capacity and Numbers of Power Stations for the North West Region of European Russia[1] (1993)

Type of Station	Hydro	Gas	Oil	Coal	Total Number of Power Stations
Capacity Group					
3000MW+	0	0	0	0	0
2000-2999 MW	0	0	1	0	1
1000-1999 MW	0	0	0	0	0
500-999 MW	0	0	1	2	3
250-499 MW	1	1	4	6	12[2]
100-249 MW	12	6	9	4	31[3]
50-99 MW	7	2	6	5	20
0-49 MW	13	3	4	5	25
Total Number of Power Stations	33 (36%)	12 (13%)	25 (27%)	22 (24%)	92 (100%)

Notes
1. Calculated from data shown in IEA (1994), pp. 212-23.
2. Including 1 peat-fired station.
3. Including 1 peat-fired station and 1 oil shale fired station.

Table 5.10 *Fuel Mix and Power Station Capacity for European Russia (1993)*

Type of Station	Hydro	Gas	Oil	Coal	Total Capacity (MW)
Capacity Group					
3000MW+	0	12,300	0	3,700[1]	16,000 (10%)
2000-2999 MW	2,500[2]	15,000[2]	2,500[2]	7,500[2]	27,500 (17%)
1000-1999MW	3,000[2]	21,000[2]	1,500[2]	4,500[2]	30,000 (19%)
500-999 MW	2,250[2]	17,250[2]	2,250[2]	6,750[2]	28,500 (18%)
250-499 MW	4,875[2]	14,250[2]	5,625[2]	8,625[2]	30,375 (19%)
100-249 MW	2,975[2]	5,775[2]	4,025[2]	4,375[2]	17,150 (11%)
50-99 MW	825[2]	1,950[2]	900[2]	1,425[2]	5,100 (3%)
0-49 MW	525[2]	1,225[2]	37[2]	350[2]	2,475 (2%)
Total Capacity (MW)	16,950	88,750	17,175	3 7,225	160,100[3] (100%)

Notes:

1. Reproduced from data provided in IEA (1994), pp. 212-23.
2. Remainder of capacity data calculated from the numbers of power stations shown for each capacity range and fuel type provided in Table 5.4, multiplied by the median of the capacity range for each row (for example, 2500 MW = 2,500MW × 1).
3. The overall total for the capacity of power stations in European Russia is provided in IEA (1994) as 164,580 MW in 1993, including 16,176MW for nuclear power stations. The calculated total capacity of 160,100 is therefore less than 8 per cent higher than the actual figure, and is therefore sufficiently accurate for the purposes of estimating the proportions of total generating capacity accounted for by power stations of defined size categories.

A re-powering programme could therefore be considered for some of these power stations, and also extended to the substitution of gas for some oil-fired and coal-fired units, because of the environmental advantages and efficiencies now available from contemporary gas turbine units (see Chapter 4). The implementation of such a re-powering programme for these small stations would depend, however, upon their location in relation to larger utilities, and as to whether a schedule of utility rationalisation and investment in distribution facilities may be more appropriate. For those power stations remote from larger utilities and having old facilities, replacement by gas turbine units may prove to be the most efficient option, provided that adequate servicing and maintenance can be provided. For very small utilities, however, which may not have the infrastructure to support gas turbine technology, gas engines may be more appropriate in view of their relative simplicity in operation and maintenance.

The opportunities provided by modern gas turbine units, should not detract from the significance of oil and coal in the firing of small units, and the importance of small hydro power units in North West Russia. Technologies are available for the low NOx combustion of fuel oil, although reductions of SOx emissions will be contingent upon further refining of this fuel. Furthermore, in those regions where coal remains as an important fuel, circulating fluidised bed (CFB) systems have now reached a stage of practical development (see Chapter 4) where they could be introduced into operation in power stations of various capacities, including less than 100MW capacity utilities in the Central, Urals and North West regions of European Russia. For European Russia as a whole, it can be concluded from the data shown in Table 5.4 that some 7,000MW capacity of coal burning capacity is installed in power stations of less than 250MW capacity, which is a capacity range suitable for the installation of fluidised beds (see Chapter 4).

Further data on Russian coal-fired power stations published in 1996 suggests that this capacity has been reduced from a 1993 level of 37,200MW in 1993 (see Table 5.10) to some 21,400MW in European Russia (Clarke, 1996, pp. 18, 19, citing IEA Coal Research, 1996).[3]   Of particular importance to this current research, however, is the data presented in that source for coal-fired power generation capacity in European Russia including the Urals (some 21,400MW, including some 9,000MW in the Urals), and Siberia and the Far East (some 27,800MW).

It is apparent, therefore, that the majority of developments related to the reduction of SOx and NOx emissions should be focused on Siberia (especially the Kuzbass, Irkutsk and Kansk-Achinsk regions accounting for some 11,400MW of coal-fired capacity) and the Urals, in order to obtain the maximum impact on national emissions; although international political factors may dictate that the installation of any such equipment occur close to the European borders (see Chapter 7). A dependence on gas as a major fuel for electricity production is also apparent for some of the other European republics constituting the former Soviet Union (IEA, 1994). In the case of Ukraine more than 70 per cent of its electricity output of 182TWh in 1992 was generated in thermal plants, and 45 per cent from gas. For Belarus, which produced some 40TWh in 1990, the power generation capacity of 7GW in 1992 was almost completely thermal, with gas accounting for some 39 per cent of the fuel consumed; whilst in Latvia in 1993 gas accounted for some 46 per cent of fuel consumed in thermal power stations (although these power stations only produced 1.3TWh in that year) and some 54 per cent in Lithuania. In every case except Ukraine, which has some indigenous supplies, these republics have relied almost exclusively on Russia as a source of supply for this fuel, as the infrastructure was previously in place for these deliveries through the integrated Soviet pipeline system.

In addition, oil has been an important fuel in Latvia (39 per cent of fuel consumed in thermal power stations in 1992) and Lithuania (45 per cent of fuel consumed in thermal power stations in 1992), and the major fuel in Belarus (58 per cent of fuel consumed in thermal power stations in 1992) (IEA 1994). Russia is also the exclusive supplier of this fuel to these republics, except for Ukraine which has some small indigenous supplies; and for the two smaller republics it can be questioned whether the costs of import of these fuels and subsequent power generation are cheaper than the import of electricity through the former Soviet national grid. Strategic issues related to power independence, however, may override those of cost.

As an alternative to continued import of Russian oil and gas, the non-Russian republics may prefer to develop alternative fuels or techniques for power generation. Coal has been an important fuel for Ukraine, previously accounting for some 34 per cent of electricity produced, and its territory contains significant coal deposits although some are of qualities with high sulphur, and more appropriate to the metallurgical industry. Oil shale has been the most important fuel in Estonia in view of its plentiful supplies in that region, accounting for some 94 per cent of the 3.3 GW capacity in 1992; but the high levels of SOx emitted from the combustion of this fuel are a major cause of ecological concern in the Scandinavian region. Latvia and Lithuania are less dependent than the other republics on conventional thermal power stations for electricity generation, as 66 per cent of Latvian electricity is generated in hydro power stations, and 78 per cent of

Lithuanian electricity is from nuclear power stations.   There is some dependence on Russia for the continued operation of these latter utilities as the reactors are of Soviet design (IEA, 1994).

FUTURE OUTPUT AND CAPACITY

Output

Various estimates exist on forecast demand for electricity by the year 2010, and these are shown in Table 5.11. From these data, it can be seen that estimates for electricity demand vary by factors of two in the years 2000 and 2010. Makarov et al.'s (1994) demand estimate is more optimistic than those of the other cited sources, and Bushuev's (1993) estimates of electricity production during 1995, 2000 and 2010, are marginally higher than those provided by Lagerev and Khanaeva (1993).   From the information given in Table 5.11, averages have been calculated for the estimates provided, to arrive at averages of 983TWh for 1995, 1000TWh for 2000, and 1293TWh for 2010. These estimates are similar to those provided by D'yakov (1995), who envisages that the present decline in Russian demand will cease in 1996-98, and that by 2005 demand will have reached the 1990 level (1082TWh) level, and exceeded this level by some 18-20 per cent (that is, 1298 Wh) in the year 2010.

Fossil Fuel Generating Capacity

A major problem facing the electricity power generation industry is that of estimating required capacity under conditions of economic uncertainty giving rise to questions over levels of demand and output, and a lack of clarity over power station closures where plant is old or demand has fallen to uneconomic levels.   Furthermore, additional information is required over practical investment opportunities in the industry (see Chapter 7), which will influence efficiencies of operation.

By the year 2000 it is estimated by D'yakov (1995) that some 25 per cent of thermal power station capacity will have been removed from operation (that is 37,000MW, assuming a Russian thermal power generating capacity of 149,000MW in 1990), and some 60 per cent (that is, 89,000MW) by 2010. Over the 15 year time interval between 1995 and 2010, D'yakov also envisages the installation of new and refurbished thermal capacity of some 60,000MW (plus 8,500MW of hydro power stations and 11,000MW of

Table 5.11 Estimates of Forecast Demand for Electricity in Russia

	1990	1995	2000	2010
Official 1[1]	100%	97%	112%	152%
Official 2[1]	100%	95%	108%	132%
Mission A[1]	100%	82%	88%	112%
Mission B[1]	100%	75%	75%	112%
Mission C[1]	100%	68%	55%	75%
Makarov et.al[2]	100%	115%-125%		
	(1082TWh)	(1250-1350Wh)		
Lagerev & Khanaeva[3]	100%	94%	102%	122%
	(1082TWh)	(1015-1020TWh)	(1105-1110TWh)	(1310-1320TWh)
Bushuev[4]	100%	96%-97%	105%-109%	126%-138%
	(1082TWh)	(1035-1050TWh)	(1135-1180TWh)	(1360-1490TWh)
Average of forecasts	100%	91%	92%	120%
	(1082TWh)	(983TWh)	(1000TWh)	(1293TWh)

Sources:

1. IEA (1994, p.194) ('Official 1' & 'Official 2' estimates were compiled by experts from the Russian Power Engineering Institute, and differ in terms of assumptions over economic growth rather than electricity intensity. The three 'mission statements' were compiled by experts from IEA and the World Bank and differed in terms of assumptions about economic growth and electricity intensity. At the time of the estimates, hard data were only available up to 1991, but subsequent data for 1992 and 1993 for GDP, industrial production and total electricity consumption tended to be more in line with Western projections than with Russian ones).

2. Makarov et al., (1994).
3. Lagerev and Khanaeva (1993).
4. Bushuev (1993).

nuclear plant) including some 22,000-25,000MW of combined cycle equipment installed at the rate of 1,000MW per annum up to the year 2000 and 2,000MW per annum between 2000 and 2010. Makarov et al. (1994) provided an estimate that thermal capacity would remain at 135,000MW from 1992 to 1995, and will then increase to 142,000MW in 2000, 154,000MW in 2005 and 161,000MW in 2010, or a net increase of some 26,000MW (or an increase of some 49,000-50,000MW if hydro and nuclear power generation capacity are also included). It is envisaged that the capacity of combined heat and power stations will increase by some 15,500MW (from 71,300MW to 86,800MW, or an increase of 21.7 per cent) and that of condensing power stations will increase by some 11,200MW (from 63,700MW to 74,900MW, or an increase of some 17.5 per cent). The net increase of 26,000MW thermal power station capacity (or 26,700MW if the increases in cogeneration and condensing capacity are totalled) is to be achieved by the construction of new power station capacity to the level of some 40,000-50,000MW, and the refurbishment of some existing 100,000MW capacity. This is a large proportion of the total thermal generating capacity inferring that only some 19,000MW will be closed down and some 16,000MW capacity retained without refurbishment, and is a far higher refurbishment figure than that of D'yakov cited above who appears to predict a much higher closure programme resulting in a thermal generating capacity of some 120,000MW in 2010. Makarov et al. envisage the construction of new combined cycle power plants in North Tyumen, North Caucasus and European Russia, and the construction of new coal-fired capacity in Siberia and the Urals. Furthermore, new power station capacity will probably also be required in the Far East to meet infrastructural requirements in that region.

Other estimates provided by Gorin and Shabalin (1995) envisage a reduction in 1994 capacity from 170,000MW in 1995 to 110,000MW in 2010, with total capacity achieving either some 180,000MW capacity or some 240,000MW in that year for either a 'low demand' or a 'high demand' scenario, respectively. These data suggest an increase of some 70,000 to 130,000MW by that year. According to their estimates provided for 1995-2005, it is likely that the majority of this increase in capacity will be achieved through investment in fossil-fuelled co-generation and condensation plants, with more than 60 per cent of these extensions being achieved through reconstruction and modernization including the installation of combined cycle equipment. By using combined cycle technologies in 120 new and modernized power plants, it is intended to introduce some 42,700MW$_t$ of generating capacity (Gorin and Shabalin,

1995), or some 24,700MW$_e$ if a cycle efficiency of 58 per cent is assumed. This estimate of the quantity of refurbished plants differs from that provided by Glebov (1995) (600 power generation sets to be refurbished), although the differences may be accounted for by the use of different terminology (that is, power generation 'plant' or power generation 'sets'). For the purposes of estimating demand for new power station equipment, therefore, it has been decided to select a figure on which Makarov and D'yakov agree, and also coincides with IEA (1994) data for plant currently over twenty years old; namely the installation of some 40,000-50,000MW in new or refurbished capacity divided equally between combined cycle and condensing power stations from 1995 to 2010. This assumption would also provide an estimate of increased combined cycle power station capacity of some 25,000MW, which is also identical to that quoted from Gorin and Shabalin in the previous paragraph, assuming a cycle efficiency of 58 per cent.

These latter estimates are less optimistic than those previously provided in 1992 (by the Institute of Energy Research of the Russian Academy of Sciences (Akimov et al., 1992), namely a combined cycle capacity of some 90,000-140,000MW in 2010 (although these figures may be for thermal capacity). These lower estimates, however, do raise questions as to the technological feasibility of the associated proposed programme, as typical maximum capacity of new combined cycle plant is some 900MW, and some 165MW for circulating fluidised beds (the preferred 'clean coal' option for coal-fired power stations). To achieve the anticipated increases in capacity, therefore, it is likely that some refurbishment will also be necessary to large gas-fired or coal fired condensing stations, which can operate steam turbines of up to 800MW to achieve power station capacities of greater than 3000MW.

Combined Heat and Power Capacity

Further developments relating to efficiency improvements, will need to take account of the widespread use of combined heat and power (CHP) systems in the former USSR. CHP power stations (*Teploelektrotsentrali*, abbreviated to *TETs*) produced approximately some 35 per cent of the former USSR's medium and low temperature heat supply in 1990 (5.5 billion GJ from a total of 16 billion GJ, assuming that the definitions of 'heat' are consistent), or some 50 per cent of heat output from centralised supplies (11 billion GJ) (Sokolov, 1993). It is likely that that proportion has continued throughout the region since that date, although the geographical spread of CHP stations throughout the former Soviet republics will be contingent upon fuel availabilities for generation of power by thermal methods, and the distribution of population densities.

The proportion of electrical power generated by CHP stations in the former USSR is less easy to estimate, however. The IEA (1994) report for Russia, for example, refers to 797TWh of electricity output from CHP power stations in that republic in 1990, which is identical to the output figure from Russian thermal power stations in that year, and almost 80 per cent of the total Russian production of electricity from all sources (1082.2TWh in 1990). Even if there was no CHP electricity production in any other republic, therefore, it is apparent that Russian CHP stations alone would have accounted for some 46 per cent of total Soviet output of electrical power (1726TWh) in 1990.

There is, however, certain evidence available to suggest that these data should be treated with caution, particularly as the IEA (1994) report itself refers to a 1990 Russian CHP capacity of 65GW within a total Russian net conventional thermal capacity of 131.6GW, and total public generating capacity of 195GW. These data therefore suggest that CHP capacity accounts for some 49 per cent of Russian thermal generating capacity, and some 33 per cent of total generating capacity. Unless the electrical efficiency or annual operating times of CHP stations are significantly higher than those of condensing power stations, it is therefore doubtful whether the proportion of electrical energy generated from CHP stations would have exceeded 33 per cent of the Russian total in 1990, or 49 per cent from thermal stations.

Furthermore, Sokolov (1993) provides a capacity figure of 63 GW in 1990 for CHP turbines in power stations responsible to the Russian Ministry of Fuels and Energy (*Mintopenergo Rossii*), 85GW for power stations responsible to the Ministry of Energy of the USSR (*Minenergo SSSR*) and 98GW for all CHP power stations in the former USSR. If the capacity of thermal Soviet power stations in 1990 is assumed to have been some 241GW in 1990 (including 149GW for the Russian Federation), and some 344GW in total for all types of power station during that year (including 213GW for the Russian Federation) (*Toplivno-Energeticheskii Kompleks*), then CHP stations would appear to have accounted for some 41 per cent of thermal power generation capacity, and some 28 per cent of capacity for all power stations; or 35 per cent and 25 per cent respectively if stations only within *Minenergo SSR* are considered, and 43 per cent and 30 per cent respectively for Russian CHP stations responsible to *Mintopenergo*. Using the same arguments as in the previous paragraph, therefore, it can be concluded that

CHP stations probably accounted for between 35-43 per cent of the output from thermal power stations and 25-30 per cent of the output from all power stations.

Other data available from Sokolov (1993) for the 1985-90 time interval states that condensing power stations accounted for some 64-66 per cent of the electrical energy output from thermal power stations administered by *Minenergo*, and CHP stations accounted for 34-36 per cent. In addition, Sokolov provides a 1990 output figure of 468TWh for electrical energy output from CHP turbines within a total output of 1198TWh for thermal power stations responsible to *Minenergo SSSR*, and an output of 515TWh from all Soviet CHP turbines within a total output of 1726TWh from all Soviet power stations. These data therefore suggest that electricity output from CHP stations has accounted for some 39 per cent of power generated by thermal power stations, and some 30 per cent of total power generated in the former USSR in 1990. CHP power stations responsible to *Mintopenergo Rossii* produced 350TWh (or 48 per cent) of the 734TWh produced by thermal power stations responsible to that organisation, or 32 per cent of total Russian electricity output (1082TWh).

It is important however to note that Sokolov's table provides a 'combined electrical energy output figure from CHP stations' (*kombinirovannaya vyrabotka elektricheskoi energii na TETs*) of 263TWh, compared to the previously cited 468TWh figure which relates to 'electroenergy output from combined heat and power turbines' (*vyrabotka elektroenergii teplofikatsionnymi turbinami*). Furthermore, Sokolov also refers to an output ratio of 80kWh/GJ from *Minenergo SSSR* CHP stations, which is consistent with a thermal energy output of 3.3GJ in 1990 for an electrical output of 263TWh. If the lower output figure is the correct estimate, therefore, then CHP stations only accounted for some 15 per cent of total electrical output in the former USSR, and some 22 per cent of output from thermal power stations.

Data from Sokolov also suggests that Soviet CHP stations were operationally more fuel efficient than their condensing counterparts (269 net grams of standard fuel per kWh compared with 326 grams per kWh), although it is not totally clear from that data whether the thermal energy generated from CHP power stations is converted to equivalent electrical energy units and included in the total energy output figure from these generating sets. This can be highly significant, however, as thermal energy accounts (on the average) for some 73-78 per cent of the total energy output from combined heat and power stations (234GW thermal capacity of CHP turbines compared with 85GW for electrical capacity, or 1 GJ [278kWh equivalent] of thermal energy per 80kWh of electrical energy).

By increasing inlet pressures into the steam turbine, from 13MPa (at 550°C) to 23MPa (540°C), to provide supercritical inlet conditions whilst maintaining steam outlet temperature at 100°C, an increase of 17 per cent can be achieved in terms of the ratio of electrical energy to thermal energy. If the supercritical steam turbine is then operated in conjunction with a gas turbine having an inlet temperature of 1000°C, this ratio is increased by a further 20 per cent (Sokolov, 1993). It is therefore apparent that particular potential exists for the transfer of gas turbine technology which has been alluded to in a previous section of this chapter. This technology is particularly relevant within the context of CHP power stations, although scope will also exist for improvements in the general efficiency of operation of steam-driven units.

Fuel Requirements

Estimated fuel requirements by Lagerev and Khanaeva (1993) are shown in Table 5.12, which also includes power and heat output, and fuel requirement estimates by Bushuev (1993) for comparative purposes. These estimates are seen to compare very closely for heat estimates, but Bushuev's fuel estimates for gas and coal exceed those for Lagerev's and Khanaeva's 'gas rich' and 'coal rich' scenarios respectively, as a consequence of the differences in estimates of power output.

Lagerev and Khanaeva also provide estimates of fuel required if output from nuclear power stations is to increase at a slower pace than electrical output as a whole, and their 'gas-rich' and 'coal-rich' forecast scenarios envisage increases in gas output for electrical power generation increasing by some 19-23 per cent in 2010 compared with 1995, and West Siberian gas increasing by some 11-17 per cent. Of more significance, however, is the anticipated increase in coal demand which is expected to increase by some 25-43 per cent overall, some 5-26 per cent in the Kuzbass region, but a staggering 118-172 per cent in the Kansk-Achinsk region. If the same proportions are used for the estimated average electricity output figure of 1293 TWh for the year 2010 calculated above it appears that gas requirements will vary between 763 and 790 billion m^3(bcm), and coal requirements between 360-410 million tons, to meet the total domestic and export demands.

Table 5.12 Estimates of Fuel Requirements for Power Generation in 1995

	1995	2010	
		'gas' option	'coal' option
Electricity production TWh	1030	1355	1355
	(100%)	(132%)	(132%)
	[1035-1050]	[1360-1490]	[1360-1490]
including nuclear power stations	115	140	140
hydro power	(100%)	(122%)	(122%)
	160	215	215
thermal power	755	1000	1000
	(100%)	(132%)	(132%)
Export of electricity, TWh	10	35	35
Centralised heat production, million	2010	2330	2330
G cals	[2100-2150]	[2200-2370]	[2200-2370]
Fuel extraction			
gas, billion m^3 (bcm)	670	830	800
	(100%)	(124%)	(119%)
gas from W. Siberia	620	725	690
	(100%)	(117%)	(111%)
coal, million tons	300	375	430
	(100%)	(125%)	(143%)
	[340-350]	[400-440]	[400-440]
from Kuzbass	95	100	120
	(100%)	(105%)	(126%)
from Kansk-Achinsk	55	120	150
	(100%)	(218%)	(273%)
Production of fuel oil, million tons	55	47	47
Export of gas, billion m^3 (bcm)	215	245	245
Export of coal, million tons	50	45	45
Total costs on new construction, (1995-2010), billion roubles, 1989 prices	-	30	31.52

Note: The data is taken from Lagerev and Khanaeva (1993) except for the estimates shown in square brackets taken from Bushuev (1993). Figures shown in round brackets are based on Lagerev and Khanaeva's data, with 1995 as the base year.

If such increases are required to meet projected 2010 levels of electricity output without a significant increase of nuclear generating capacity, it is apparent that significant developments will be required in gas combustion to increase efficiency, and advanced coal technologies for the cleaner combustion of both Kuzbass and Kansk-Achinsk fuel (see Chapter 4). It is important to note that the power generation sector consumed some 28 per cent of gas produced in 1990 (179.0bcm within a total of 640bcm) and some 27 per cent in 1993 (165.4 bcm within a total of 618bcm) (Stern, 1995, p.33). Future scenarios envisage the power generation sector consuming some 22-38 per cent of production in the year 2000, using Stern's quoted estimates of 244-248bcm provided by the All-Union Scientific-Research Institute for Gas (VNIIGAZ) for the year 2000, an estimate of 167bcm by Gazprom, and 140bcm based on an estimate of 30 per cent decline in the power generation sector compared with 1990 (Stern, 1995, pp. 41, 42, 85), and a production level of 640bcm.

Very few of the remaining former Soviet republics, however, have the same degree of choice of fuels policy as Russia, as they have far lower levels of energy resources. Each of these republics are consequently attempting to economise on oil and gas consumption to reduce their dependence on Russian imports of these commodities, whilst also attempting to develop their own indigeneous energy resources (for example, coal in Ukraine and parts of Central Asia, hydro-power in Latvia and Kirghizia, oil and gas reserves in Ukraine, Azerbaijan and central Asia; and the continued use of nuclear power in most of the European regions of the former USSR). The development of these reserves will obviously take several years, however, which provides limited options for fuels policy in these newly independent states, and it is useful to review their possibilities for the use of hydroelectricity in view of its complete absence of atmospheric pollution. The following section will consequently provide a summary of major proposals for hydro-electricity generation in the former USSR.

Hydro-electricity Generating Capacity

In 1992, hydro-power accounted for some 20 per cent of Russian electricity generation capacity (43,000MW) and some 17 per cent of output (172TWh) (IEA, 1994, pp. 208, 211). The development of hydro-electric power stations in Siberia could add to the present high levels of hydro-energy capacity within the region (some 12,421MW within a national total of 27,161MW). The hydro-project at Boguchanskoi for 3,000MW is at an advanced state of construction and a further 3,600MW is planned for extensions to the existing Siberian network over the next decade. A further 3,600MW is also planned for north Caucasus, out of a total additional planned hydro capacity of 7,334MW. A theoretical potential of 16,000MW is claimed for Russia as a whole, but additional transmission capacity would also be required to utilise that additional generating capacity (IEA, 1994, pp. 199, 200), and D'yakov (1995) provides an estimate of 8,500MW of new capacity in addition to 11,000MW in nuclear plant to meet shortfalls in thermal and hydroelectricity capacity

In Ukraine, the proportion of hydroelectric capacity has been some 4,700MW, but this only represents some 9 per cent of total capacity and provides some 3 per cent of output (11.9TWh) and Belarus, Lithuania, Estonia and Moldova have even smaller proportions of hydro-generation capacity within their total resources (IEA, 1994, pp. 45-252). In Latvia, however, hydro power accounts for some 75 per cent of capacity (almost 1,500MW), although the output from these plants as a proportion of total output can vary according to rainfall (IEA, 1994, pp. 199, 200). Furthermore, scope exists for the development of a further 113MW additional capacity in Latvia, which already has a significant proportion of its electricity generating capacity located in three hydro plants at Riga, Kegum and Plavinas (IEA, 1994, p. 113). Investment in each of these proposals, however, would require some care to be exercised over the ecological balance in the relevant region. Hydro electricity is currently generated in some 90 per cent of the power stations in Kirghizia, and those stations account for almost 80 per cent of the republic's generating capacity (2,700MW within a total of 3,400MW). Furthermore, some 10 per cent of generated power is exported, and there is potential for further development with the adjacent republics of Uzbekistan, Tadzhikistan and Turkmenistan, as well as southern regions of Kazakhstan and the Sintszyan-Uiaursk autonomous region of that republic. Current and proposed projects include the enlargement of the Tash-Kumyrskii GES and Kambatinskii GES-2; and the commencement of construction of 1,900MW capacity at the Kambaratinskoe GES 1 (Tuleberdiev, 1994). In addition to the development of capacity for export, such projects would also reduce the necessity of imports of oil, gas and coal in the Central Asian republics.

CONCLUSIONS

This chapter has provided information on the capacity and fuel mix of power stations in the former USSR using data available for the early and mid 1990s, and plans for future investment in the industry. The information presented has demonstrated the dominant position held by gas as a fuel for power generation in the former USSR as a whole, and in Russia in particular, with consequent preferences for the future development of gas turbine and combined cycle technologies for power generation. The data has also demonstrated the continuing importance of coal in several regions of Russia, particularly in the Urals and Siberia. There is consequently a case for the further development of clean coal technologies for implementation in those regions.

The progress that has occurred in the transfer of these technologies is described in Chapter 6; and further information on the political, economic and commercial factors which have influenced these transfers is provided in Chapter 7. The data provided on future plans clearly require future monitoring and evaluation, however, in view of the continuing political and economic crises in the former USSR; and further clarification is needed as to whether some of the estimates presented refer to thermal or electrical capacities.

NOTES

1. *Toplivno-Energeticheskii Kompleks*, compiled by the All-Russian Institute of Integrated Fuel and Energy Problems of the Russian Ministry of Economics, and published by the Russian Federal Service of Geodesy and Cartography in 1993.
2. Estimated from the capacities provided for Belarus (7010MW in 1992), Estonia (3422MW in 1990), Latvia (2033MW in 1990), Lithuania (5690MW in 1992), Moldova (2998MW in 1993) and Ukraine (53569MW in 1993) see IEA (1994), pp.. 35-49, 81-91, 111-51, 239-55.
3. Although the IEA (1994) data refers to power stations of greater than 10MW, and the IEA (1996) data refers to power stations of greater than 50MWe, the data on capacities are comparable as there were no coal-fired power stations of less than 50MW listed in the 1994 publication.

REFERENCES

Akimov, N.K., Berezinets, P.A., Vasilev, M.K., Alekso, A.I., Gvozdev, A.V., Ershov, Yu.A., Ol'khovskii, D.D., Petrov, Yu.V., Proskuryakov, G.V., Barinberg, G.D. and Chukanov, A.D. (1992), 'Teplofikatsionnaya parogazovaya ustanovka moshnostyu 130MWt', *Teploenergetika*, No. 9, pp. 22-7.

Baranovskii, A.I. (1993), 'Formirovanie struktury upravleniya i elektroenergicheskogo rynka v Rossii', *Elektricheskie stantsii*, No. 7, pp. 2-6.

Bushuev, V.V. (1993), 'Novaya energeticheskaya politika Rossii (osnovnye polozheniya kontseptsii)', *Energetik*, No. 5, pp. 4-9.

Clarke, L.B. (1996), *Coal Prospects in Russia Perspective* (IEAPER 27), London: IEA Coal Research.

Cooper, J.M. (1991), *The Soviet Defence Industry: Conversion and Reform*, London: RIIA/Pinter.

D'yakov, A.F. (1993), 'Reformy upravleniya elektroenergetikoi Rossii', *Energetik*, No. 12 pp. 3,4.

D'yakov, A.F. (1995), 'Elektroenergetika Rossii i energeticheskaya bezopasnost'', *Energetik*, No. 10, pp. 2-4.

Elektricheskie Stantsii (1993), No. 3, pp. 1-9 (Editorial).

Energetik (1994) No. 10, pp. 2, 3 (A. F. D'yakov).

Glebov, V.P. (1995), 'Ekologicheskaya bezopastnost', teploenergetika i investitsii, *Energetik*, No. 11, p. 2.

Gorin, V.I. and Shabalin, V.L. (1995), 'Investment possibilities in Russian Power Industry', paper presented at a Pre-conference Seminar at the 1995 Powergen Exhibition (Amsterdam).

Hewett, E. (1984), *Energy, Economics and Foreign Policy in the Soviet Union*, Washington DC: Brookings Institution.

International Energy Agency (IEA) (1994), *Electricity in European Economies in Transition*, Paris: OECD.

International Energy Agency (IEA) Coal Research (1996), *Coalpower database*, London: IEA Coal Research.

Kushnarev, F.A. (1994), 'Aspekty re-organizatsii sistemy Rostovenergo', *Elektricheskie stantsii*, No. 6, pp. 16-19.

Lagerev, A.V. and Khanaeva, V.N., (1993) 'Stsenarii razvitiya Energeticheskogo Kompleksa Rossii', *Energetik*, 1993, No. 11, pp. 5-8.

Makarov, A.A., Bruilov, V.P., Volkova, E.A. and Vol'kenau, I.M. (1994), 'Vosprosy ekonomicheskoi otsenki posledstvii dosrochnogo zakrytii AES', *Energetik*, No. 7, pp. 5-8.

Red'ko, V.I. (1993), 'Privatizatsiya regional' nykh elektricheskikh kompanii', *Elektricheskie stantsii*, No. 7, pp. 14-19.

Rosengaus, J. (1986), 'The Selection of Steam Parameters for Soviet Thermal Power Plants', in Young, K.E. (ed.), *Decision Making in the Soviet Energy Industry: Selected Papers with Analysis*, Delphic Associates, Falls Church, VA, pp. 1-31.

Rosengaus, J. (1987), *Soviet Steam Generator Technology : Fossil Fuel and Nuclear Power Plants*, Delphic Associates, Falls Church, VA, pp. 1-51.

Sokolov, E.Ya. (1993), 'Razvitie teplofikatsii v Rossii', *Teploenergetika*, No. 12, pp. 2-7.

Stern, J.P. (1995), *The Russian Natural Gas 'Bubble': Consequences for European Gas Markets,* London: The Royal Institute for International Affairs: Energy and Environmental Programme.

Tuleberdiev, Zh.T. (1994), 'Sostoyanie i perspektivy elektroenergetiki v Kirgyzkoi Respublike', *Energetik*, No. 5, pp. 3, 4.

Whitefield, S. (1993), *Industrial Power and the Soviet State*, Oxford: Clarendon Press.

Zagyanskii, A.I. and Red'ko, V.I. (1994), 'Privatizatsiya v elektroenergetike: problemy i resheniya', *Elektricheskie stantsii*, No. 5, pp. 12, 13.

Table 5.1A Location, Capacity, Installation Dates, and Fuel Mix of Power Stations in the Former USSR Power Stations of 3000+MW Capacity

Power Station	Installation Dates	Average Age (Yrs)	Total Capacity (MW)	Fuel Mix (%)	Gas-fired Capacity (MW)	Coal-fired Capacity (MW)	Oil-fired Capacity (MW)
Surgutskaya GRES 2	1985-88	9	4,800	Gas 100	4,800	-	-
Ekibastuzskaya GRES 1	1980-84	13	4,000	Coal 96 Oil 4	-	3,840	160
Reftinskaya GRES	1977-80	17	3,800	Coal 99 Oil 1	-	3,762	38
Zaporozhskaya GRES	1975-77	19	3,600	Oil 54 Coal 46	-	1,656	1,944
Kostromskaya GRES	1980	15	3,600	Gas 92 Oil 8	3,312	-	288
Uglegorskaya GRES	1975-77	19	3,600	Oil 83 Gas 17	612	-	2,988
Surgutskaya GRES 1	1972-83	18	3,464	Gas 100	3,464	-	-
Krivorozhskaya GRES	1965-73	26	3,000	Coal 73 Gas 15 Oil 12	450	2,190	360
Syrdarinskaya GRES	1972-81	19	3,000	Gas 80 Oil 20	2,400	-	600
Total Capacity			32,864		15,038	11,448	6,378

Table 5.1A continued
Power Stations of 2,000-2,999MW Capacity

Power Station	Installation Dates	Average Age (Yrs)	Total Capacity (MW)	Fuel Mix (%)	Gas-fired Capacity (MW)	Coal-fired Capacity (MW)	Oil-fired Capacity (MW)
Burshtynskaya GRES	1965-69	28	2,976	Coal 84 Oil 16	-	2,500	476
Ryazanskaya GRES	1973-74	22	2,800	Coal 79 Gas 21	588	2,212	-
Ermakovskaya GRES	1968-75	25	2,400	Coal 99 Oil 1	-	2,376	24
Zainskaya GRES	1963-72	28	2,400	Gas 96 Oil 4	2,304	-	96
Iriklinskaya	1970-72	24	2,400	Gas 91 Oil 9	2,184	-	216
Konakovskaya GRES	1965-69	28	2,400	Gas 82 Oil 18	1,968	-	432
Lukoml'skaya GRES	1969-74	24	2,400	Oil 100	-	-	2,400
Stavropol'skaya GRES	1974-83	17	2,400	Gas 73 Oil 27	1,752	-	648
Troitskaya GRES	1965-87	29	2,200	Coal 97 Oil 3	-	2,134	66
Kirishskaya GRES (Lenenergo)	1969-75	23	2,120	Oil 100	-	-	2,120

Azerbaidzhanskaya GRES	1981-89	10	2,100	Gas 58 Oil 42	1,218	-	882
Moldavskaya GRES	1964-74	26	2,020	Coal 60 Oil 26 Gas 15	283	1,212	525
Zmievskaya GRES	1960-64	33	2,000	Coal 64 Gas 22 Oil 14	440	1,280	280
Starobezhevskaya GRES	1961-67	31	2,000	Coal 42 Oil 40 Gas 18	360	840	800
Total Capacity			32,616		11,097	12,584	8,965

Table 5.1A continued
Power Stations of 1,000-1999MW Capacity

Power Station	Installation Dates	Average Age (Yrs)	Total Capacity (MW)	Fuel Mix (%)	Gas-fired Capacity (MW)	Coal-fired Capacity (MW)	Oil-fired Capacity (MW)
Tashkentskaya GRES	1963-71	28	1,920	Gas 74 Oil 26	1,421	-	499
Mosenergo GRES 4 (Shaturskaya)			(1912)				
Karmanovskaya GRES	1968-73	25	1,800	Gas 96 Oil 4	1,728	-	72
Ladyzhuskaya GRES	1970-71	25	1,800	Gas 48 Coal 41 Oil 11	864	738	198
Litovskaya GRES	1967-72	25	1,800	Oil 73 Gas 27	486		1,314
Pridneprovskaya GRES	1963-66	31	1,800	1200 Coal 50 Gas 27 Oil 12 _600_ Coal 56 Gas 32 Oil 12	324 192	600 336	276 72

Station	Period		Fuel (%)				
Slavyanskaya GRES	1967-71	26	Gas 82 / Coal 14 / Oil 4	1,780	1,460	249	71
Maryiiskaya GRES	1973-87	15	Gas 100	1,685	1,685	-	-
Estonskaya GRES	1969-73	25	Slates 98 / Oil 2	1,610	-	1,578	32
Permskaya GRES	1986-87	9	Gas 100	1,600	1,600	-	-
Voroshilovgradskaya GRES	1961-69	30	Coal 42 / Oil 37 / Gas 21	1,500	315	630	555
Cherepetskaya GRES	1963-64	32	Coal 88 / Oil 12	1,500	-	1,320	180
Pribaltiskaya GRES	1963-65	32	Slates 100	1,435		1,435	
Nazarovskaya GRES	1961-65	32	Brown Coal 100	1,400		1,435	
Krasnoyarskaya GRES	1961-65	32	Coal 100	1,350		1,350	
Primorskaya GRES	1963-65	31	Coal 93 Oil 7	1,280		1,190	
Tom'-Usinskaya GRES	1963-65	31	Coal 98 Oil 2	1,276		1,250	
Tbilisskaya GRES	1963-72	28	Oil 60 / Gas 40	1,250	500		750

Table 5.1A continued
Power Stations of 1,000-1,999MW continued

Power Station	Installation Dates	Average Age (Yrs)	Total Capacity (MW)	Fuel Mix (%)	Gas-fired Capacity (MW)	Coal-fired Capacity (MW)	Oil-fired Capacity (MW)
Belovskaya GRES	1964-68	29	1,230	Coal 98 Oil 2		1,205	25
Dzhambulskaya GRES	1967-76	24	1,230	Oil 58 Gas 42	517		713
Sredneuralskaya GRES	1969-70	25	1,224	Gas 95 Oil 5	1,163		61
Zuevskaya GRES 2	1982-88	10	1,200	Gas 74 Coal 23 Oil 3	888	276	36
Navoiiskaya GRES	1980-81	15	1,200	Gas 100	1,200		
Novocherkasskaya GRES	1965-72	27	1,200	Coal 70 Oil 21 Gas 9	108	840	252
Novoangrenskaya GRES	1985-88	9	1,200	Gas 54 Coal 43 Oil 3	648	516	36
Irkutskaya TETS10	1960-62	34	1,110	Coal 100		1,110	
Ali-Bairamalinskaya TETS2	1962-68	30	1,100	Oil 69 Gas 31	341		759

	Installation Dates	Average Age (Yrs)	Total Capacity (MW)	Fuel Mix (%)	Gas-fired Capacity (MW)	Coal-fired Capacity (MW)	Oil-fired Capacity (MW)
GRES5 Mosenergo (Shaturskaya)	1971-78	21	1,090	Gas 55 Oil 26 Peat 19	600	207	283
Gusinoozerskaya GRES	1976-88	13	1,050	Coal 85 Oil 15		893	157
Razdanskaya GRES	1971-74	23	1,110	Oil 67 Gas 33	366		744
Kurokhovskaya GRES	1972-75	22	1,460	Coal 72 Oil 28		1,051	409
Total Capacity			42,190		16,406	18,174	7,610

Power Stations of 0-999MW Capacity

Power Station	Installation Dates	Average Age (Yrs)	Total Capacity (MW)	Fuel Mix (%)	Gas-fired Capacity (MW)	Coal-fired Capacity (MW)	Oil-fired Capacity (MW)
Berezovskaya GRES	1961-67	31	920	Gas 53 Oil 47	488	-	432
Pecherskaya GRES 1	1979-87	12	840	Gas 100	840	-	
Berezovskaya GRES 1	1988	7	800	Coal 90 Oil 10	-	720	80
Shchkinskaya GRES	1964-65	31	780	Gas 100	780	-	
Yuzhnoural'skaya GRES	1961	34	770	Gas 100	770	-	
Verkhnetagil'skaya GRES	1961-64	33	770	Gas 86 Coal 13 Oil 2	662	100	8

Table 5.1A continued
Power Stations of 0-999MW continued

Power Station	Installation Dates	Average Age (Yrs)	Total Capacity (MW)	Fuel Mix (%)	Gas-fired Capacity (MW)	Oil-fired Capacity (MW)	Oil-fired Capacity (MW)
Yuzhnaya TETS Lenenergo			750				
Nevennomysskaya GRES	1964-70	28	630	Gas 60 Oil 40	378	-	252
Smolenskaya GRES	1977-85	14	630	Gas 79 Coal 21	498		132
Cherepovetskaya GRES	1976-78	18	630	Coal 96 Oil 4		605	25
Krasnodarskaya TETS	1963-64	32	610	Gas 72	440	85	85
Dobroyvorskaya GRES	1963-64	32	600	Gas 99 Oil 1	594		6
Krasnovodskaya TETS2	1984-86	32	600	Gas 100	590		
Erevanskaya TETS	1965-66	30	550	Gas 72 Oil 28	396		154
Neryungrinskaya GRES	1983	12	390	Coal 94 Oil 6		367	23
Yavanskaya TETS	1963-65	180	180	Gas 80 Coal 20	144	36	
Severnaya GRES	1960	35	150	Gas 88 Oil 12	132		18
Total Capacity			9,840		6,704	1,913	1,223

Table 5.2A Comparisons of Power Station Capacities

Power Stations of 3,000+MW

Power Station	Capacity listed in Table 5.1A (MW)	Capacity listed in *Teplo-energeticheskii Kompleks* (1993) (MW)	Capacity listed in IEA (1994) (MW)
Surgutskaya[1] GRES 2	4,800	8,000 (including GRES 1)	5600
Ekibastuzskaya[2] GRES1	4,000	4,020	
Reftinskaya[1] GRES	3,800	3,800	3690
Zaporozhskaya[3] RES	3,600	3,600	3600
Kostromskaya[1] GRES	3,600	4,200	2160
Uglegorskaya[3] GRES	3,600	5,100 (complex of 3 power stations)	3600
Surgutskaya[1] GRES 1	3,464	8,000 (inlcuding GRES 2)	3324
Syrdarinskaya[4] GRES	3,000	3,000	
Krivorozhskaya[3] GRES	3,000	2,800	3000
Total Capacity (fomer USSR)	32,864	34,520	
Total Capacity (European region of former USSR)	25,864	27,500	24,974

Notes
1. Located in Russian Federation.
2. Located in Kazakhstan.
3. Located in Ukraine.
4. Located in Uzbekistan.

Table 5.2A continued

Power Stations of 2,000 - 2,999MW

Power Station	Capacity listed in Table 5.1A (MW)	Capacity listed in *Teploenergeticheskii Kompleks*(1993) (MW)	Capacity listed in IEA (1994) (MW)
Burshtynskaya GRES[1]	2,976	2,400	2,400
Ryazanskaya GRES[2]	2,800	2,800	1,980
Ermakovskaya GRES[3]	2,400	2,400 (Ermak or Pavlodar)	
Zainskaya GRES[2]	2,400	2,400	2,400
Iriklinskaya[2]	2,400	2,400	2,400
Konakovskaya GRES[2]	2,400	2,400	2,400
Lukoml'skaya GRES[4]	2,400	2,400 (Novolukoml')	2,400
Stavropol'skaya GRES[2]	2,400	2,400(Kropotkin)	2,400
Troitskaya GRES[2]	2,200	2,460	2,337
Kirishskaya GRES[2]	2,120	2,120	2,350
Azerbaidzhanskaya GRES[5]	2,100	1,500	
Moldavskaya GRES[6]	2,020	2,500	2,520
Zmievskaya GRES[1]	2,000	2,200	2,200
Starobezhevskaya GRES[1]	2,000	1,750	2,000
Total Capacity (former USSR)	32,616	32,130	
Total Capacity (European regions of former USSR)	28,116	28,230	27,787

Notes
1. Located in Ukraine.
2. Located in Russian Federation.
3. Located in Kazakhstan.
4. Located in Belarus.
5. Located in Azerbaijan.
6. Located in Moldava.

Table 5.2A continued

Power Stations of Less than 2000MW

Power Station	Capacity Listed in Table 5.1A (MW)	Capacity Listed in *Teploenergeticheskii Kompleks (1993)* (MW)
Tashkentskaya	1920	1890
Permskaya	1600	2400
Krasnoyarskaya	1350	2270
Primorskaya	1280	1500
Tbilisi	1250	1400
Dzhambul	1230	1290
Navoiiskaya	1200	1250
Novocherkasskaya	1200	2400
Irkutskaya	1100	660
Ali-Bairamalinskaya	1100	1100
Gusinoozersk	1050	1050
Razdanskaya	1110	670
Total Capacity	15,290	17,780

6. Technology Transfer for Reduced Environmental Pollution in the Former USSR- the Roles of Western and Russian Companies

INTRODUCTION

Chapters 1 and 2 of this book have provided information to show that the aggregate levels of SOx and NOx emissions from the former USSR have been high when compared with other European countries, and that the power generation industry has been a major contributer to these emissions. Chapters 3 and 4 have described the various technological options available for the reduction of atmospheric pollution within the context of fuel availabilities and combustion systems, and have highlighted those technologies for which there is a potential for transfer to the former USSR. Chapter 5 has provided further information related to the selection of potential technologies for transfer through an analysis of the capacities and fuel mix in Russia and other former Soviet republics, drawing attention to candidate power stations for refurbishment.. This chapter therefore extends that material by focusing on the practicalities of transfer of available technologies.

This practically oriented research has been achieved through case studies of eleven Western companies engaged in the transfer of technologies to the former USSR. These case studies of Western companies have been supplemented by similar surveys of three Russian power engineering organisations engaged in the assimilation of Western power engineering technologies which have a potential for reducing the levels of atmospheric pollution from power generation in the former USSR. The information relating to those companies was obtained from published information and press reports, and visits for discussions with senior executives between 1994 and 199 using the structured questionnaire format shown in the appendix to this chapter.

WESTERN PARTICIPANTS

Generation Utilities

Every Western country has its own national power generating utilities, which may be state-owned or private, but working within a government regulated framework of safety and competition, and permitted limits for atmospheric pollution. Many Western generating companies which were previously state-owned have recently been privatized, and are now operating in a competitive environment to market their technological expertise as well as electrical power. This technical expertise includes operational procedures to improve power station efficiency, and research into the combustion process.

In view of the high levels of pollution from some plants in the former USSR, it is apparent that a demand exists for the transfer of technology related to reduced environmental pollution. In order to convert this demand into a market, however, particularly as there is a shortage of domestic investment funds in the former USSR, further exports are required of electricity and other locally manufactured or extracted products to generate the required financial resources (see Chapter 7). As Finland is by far the largest Western importer of electrical power from the former USSR, accounting for almost all of OECD imports of electric current from that region in 1992, Finnish companies have established an expertise in the purchase of this commodity. Information on the development of the power generation industry in the former USSR was consequently sought from IVO International, of Helsinki. The information gained from interviews with senior executives from that company are as follows.

IVO International Ltd

IVO International Ltd is a subsidiary of the Imatran Voima Oy (IVO) Group of companies which operates in the Finnish and international energy markets. IVO supplied some 30,372GWh of electricity to the Finnish market in 1993, of which some 15,891GWh was generated domestically and some 4,608GWh imported from Russia (IVO, 1994).

IVO International is engaged in a number of technology transfer and investment activities in the former USSR, of which the most well-known to date is the leadership of the PSI Consortium (which includes the Polar Corporation and Siemens AG as the other two partners). The agreement was concluded in the summer of 1993 with Technopromexport, and is for the delivery of a natural gas-fired power plant (*Severno-Zapadnaya*) in the city of St Petersburg to be used by the energy board for that locality (*Lenenergo*). The project is worth some US$ 330 million of which IVO International's contribution is placed at $150 million[1] for the management of the PSI Consortium as well as the supply of the main buildings and automation (IVO International, 1993, p.34).

The total capacity of the plant will be some 900MW of electrical energy and some 800MW of thermal energy, comprising two power plants each of 450MWe and 400MWth. Each power plant is constituted of two gas turbines, one steam turbine and two heat recovery steam generators. According to the project schedule, the first unit is due for completion in early 1997 followed by the second unit in 1998. The project is financed by countertrade.

This project was selected for a number of reasons, including the existence of high density domestic housing near to the power plant containing some 800,000 people, for which both electricity and heating were required in the context of some uncertainty over the safety of some of the nuclear power stations in the Leningrad region. Furthermore, as Russian industrial gas prices approach those to be found on the international market, combined cycle systems with co-generation offer higher savings from increased efficiences, compared with the direct use of gas turbine exhaust gases for temperature raising of water used in district heating.

In addition, IVO International has invested in the local production of some 130 km/yr of polyurethane - insulated large-bore district heating pipes, having a 50 year lifetime. The present pipes corrode after 10 years, and approximately 300 km of district heating pipes have to be replaced every year within the St. Petersburg network of 6,000 km.

These two large ventures are based on many years of company experience in the Russian market, in which high levels of countertrade and barter have been gradually attained. The company has also been involved in the Estonian market, with investment in pollution reduction equipment in a heavy fuel oil fired power generation plant in Tallinn. The project was supported by some Finnish government funding as it was considered cost effective to invest in this plant to reduce the level of sulphur deposited on Finland.

Fuel Companies

Coal mining, and oil and gas extraction and distribution facilities have a range of technical competences covering generating efficiency and combustion technology, and power station operation and management. An example of Western participation in the fuels and energy sectors in the former USSR is provided by the experience of the British Coal Research Establishment, which is described below.

CRE Group Ltd

CRE Group Ltd is a leading technical development and consultancy organization in the field of solid fuel combustion. The group is a private company formed from the former Coal Research Establishment of British Coal. As such, the company is the legatee of British Coal Corporation's technical know-how, and much of the intellectual property related to coal use and its environmental impacts. In recent years, the international operations of the establishment have been substantially developed, and four major projects are at various stages of development in the former USSR, namely three each in Russia and Ukraine, two in Kazakhstan, and one each in Estonia and Georgia.

The Russian Projects The two projects in Russia cover coal quality assurance and energy efficiency, with funding provided by the UK Know-How Fund (KHF) for the first project, and from the European Union's Technical Assistance for the Commonwealth of Independent States (TACIS) for the second. The study on coal quality for the Russian and coal industry (*Rosugol'*) was carried out by visits to the Kuzbass and Kansk-Achinsk coalfields, and discussions with major coal customers, including power stations responsible to the Kuzbass regional authority for the generation and distribution of electrical power (*Kuzbassenergo*), and coal users in the Urals region around Ekaterinburg.

The project was designed to identify and assist with the introduction of improved coal preparation techniques, alongside an increase in the overall efficiency of coal use in Russia. Particular attention was paid to the coal production associations being given a better understanding of serving customer needs in a competitive market, and the provision of advice on the adjustment of coal quality to meet future market needs. This advice was linked to assistance in the preparation of a case for investment in coal preparation equipment and coal use technology, which should lead to improved coal sales turnover at selected mines through the sale of a better quality product to both domestic and overseas markets. Furthermore, there is also potential for transfer of technology of coal improvement and grading equipment, either at the coal mines themselves, at marshalling and storage points, or on the customers' premises. CRE was a member of a consortium led by International Economic and Energy Consultants (IEEC) for this project. Other members of the consortium included Babcock Energy Ltd, JMC Mining Services Ltd, and Coopers and Lybrand.

The energy efficiency project is related to solid fuel heating of residential and public buildings, through the improvement of solid fuels combustion technology. The study was centred on Novgorod and Tula, and involved a survey of energy uses in the residential and public buildings sectors; assessment of the types and performances of existing solid-fuel combustion equipment; evaluation of existing heat distribution, heat controls and buildings insulation; and identification of suitable Russian manufacturers for new combustion and ancillary equipment.

A demonstration project is commencing which involves the installation of several types of efficient solid-fuel combustion appliances in residential and public buildings. These appliances range from relatively small (6-10kW) domestic downburning stoves for use with wood, peat and coal, to appliances of 1MW output to supply heating to public buildings such as schools and apartment blocks. The Western appliances have the characteristics of high thermal efficiency and clean burning, performing well with Russian bituminous coals. The following stages of this development will consist of licensed manufacture in Russian factories of Western-designed products included in the demonstration project, the installation and assessment of the appliances in a range of buildings, and the construction and evaluation of performance of low energy houses built to Western standards and heated by the new appliances. CRE was the lead contractor for this project, with the Danish Technological Institute and Deutsche Montan Technologie forming the other members of the consortium.

The Kazakhstan Power Project　This KHF funded project, which also included Powergen plc, Babcock Energy Limited and International Economic and Energy Consultants (IEEC) as members of the CRE-led consortium, is concerned with the quality, economic and environmental issues associated with the use of indigenous coal from Ekibastuz and Karaganda. The particular terms of reference of the project have been the detailed review of power generation needs, coal availabilities and coal fired plant performance; and the technical viability of emission control system retrofit including FGD and low NOx burners. A pre-feasibility study of retrofitting a pressurised fluidised bed combustion unit to an existing power plant boiler is also being carried out.

The project has consequently provided a substantial transfer of know-how for retrofit and new installation of emission control and clean combustion technologies. Furthermore, the execution of the feasibility studies will form a basis for informed identification of future investment projects. The project also led to an Asian Development Bank proposal to invest in a circulating fluidised bed combusion (CFBC) retro-fit at a power station in Alm Aty. CRE have just completed the preliminary air quality study in Alm Aty for the Asian Development Bank, and tenders will soon be invited for the power station project preparation.

The Ukrainian Project This TACIS funded project is concerned with the introduction of clean combustion technology into Ukraine against a background of 70 per cent of the existing coal fired plant approaching the end of its working life in the near future, compounded by a decrease in coal quality as many of the previously used coal resources have become exhausted. The other members of the CRE-led consortium are Lurgi Lentjes Babcock Energietechnik (LLB) and Siemens AG.

CRE has particular expertise in the development of circulating fluidised bed combustion (CFBC) technologies, and a joint European and Ukrainian team has been established to provide skills in coal utilisation with particular reference to CFBC specification; design, management and operation of power plant; power station management; and business management and investment appraisal.

The project involved detailed assessments of two proposed CFBC plant sites, evaluating and training local specialists in CFBC technical developments and operational experience, and advising on strategy for wider scale implementation of CFBC technology in Ukraine. Initial testing was carried out using Donbass anthracite on a 50kW test rig at the Institute of Energy Conservation Problems of the Ukrainian Academy of Sciences in Kiev, before subsequent development at a 1MW test rig in Frankfurt. This latter rig provides data of sufficient reliability for scaling up to a 200MW installation. Potential sources of domestic manufacture are also being sought in Ukraine, although difficulties are being encountered because of the absence of any indigenous boiler-making capacity in that republic. Surveys are consequently being made of Ukrainian heavy engineering factories having the technical capability to assimilate CFBC technology into their product profile.

CRE was of the view that the barriers to transfer of their combustion technologies were legal, commercial and financial rather than technical. Sufficient financial resources for investment in clean combustion technologies were not available in the former USSR itself, whilst World Bank and European Bank funding is related to a defined 'least cost' power generation system in the former USSR. This 'least cost' system is considered to involve a small rehabilitation programme for power stations rather than investment in new technologies such as CFBC.

Power Engineering Companies

Introduction
A summary of the major characteristics of Western power equipment manufacturing companies is shown in Table 6.1, using information available from company reports and press statements at the time of the commencement of the research described in this chapter. These published data have been supplemented by information gained from interviews, which is presented in the following sub-sections of this chapter.

Asea Brown Boveri (ABB)
From the data shown in Table 6.1, Asea Brown Boveri (ABB) emerges as the largest power engineering company in terms of value of output and breadth of product range including both boilers and turbines, in nuclear energy, and fossil fuel generation. ABB also possesses technical expertise in low NOx burner design and manufacture, particularly through Combustion Engineering (CE) of Tennessee which had also developed close business contacts with processing industries in the former USSR[2] before being absorbed into the ABB group. ABB has invested in general power engineering facilities in Poland, Hungary and Romania, boiler and turbine production facilities in the Czech Republic (PBS-Brno), and is also currently investing in power engineering development and manufacturing facilities in Russia through 'Uniturbo' (Moscow) and 'Nevsky Zavod' (St. Petersburg) joint ventures. This latter investment is reported to include the joint production of gas turbines up to 100MW capacity (*Energetik*, 1994). ABB's investments in Eastern Europe and the former USSR have been estimated at $300m (*Fortune*, 1994), whilst the company's 1993 accounts give the company's turnover in the region at $465m. ABB is therefore in a particularly advantageous position to bid for power engineering contracts in the former USSR, including low NOx equipment for coal-fired power stations because of its capability to deliver from comparatively low cost locations in close proximity to Russia, and experience in using brown coal (which is used extensively in the former USSR) as a fuel.

The company is also engaged in a number of environmental projects through its ABB Flägt subsidiary for both the power generation and metallurgical sectors, including the installation of low NOx burners in boilers in operation in a steel plant at Cherepovets. In addition, the company is engaged in the installation of filters for the Moscow electricity authority (*Mosenergo*). The business opportunities in the power generation sector, however, appear to be more difficult to fund than in the ferrous and non-ferrous metals sectors, as pollution in these latter sectors appears to be more concentrated and easier to recognise. Furthermore, the costs of the installation can be more readily financed by sales to the West of ferrous and non-ferrous metals. The company is assisted in the development of countertrade opportunities by its financial services division, which has a turnover of some $600m.

In order to deliver filtering and other environmental cleaning equipment in an economic fashion, ABB has been recently evaluating Russian factories to carry out the necessary steelwork, fabrication and construction engineering, with automation and control equipment being delivered from the West. The company's office in Moscow can carry out the requisite commercial tasks to carry out a manufacturing audit and allocate production tasks to the Russian plants, whilst the manufacture of the products and components in Russia is monitored by a team located in the factory.

Siemens Kraftwerkunion (KWU)

Siemens is a long established engineering company, which has a significant market share in the international steam and gas turbine markets, and substantial expertise in nuclear energy, achieving a 1993 turnover of some $5.3bn from its 'Kraftwerkunion' (KWU) power engineering subsidiary (see Table 6.1). The company has also made a substantial investment in the Skoda Energy Division based in the Czech Republic, and has signed a joint venture agreement with the Kalushsky Turbine Factory (Kaluga) and with the Leningrad Metals Factory (*LMZ*). This latter joint venture ('Interturbo') is currently producing turbines for installation in a combined cycle power station (*Severno-Zapadnaya*) in St. Petersburg (see IVO case above). The company's investment in the former socialist countries of Eastern Europe is therefore diverse, accounting in total for some 29 joint ventures, and the company's income from the region of $730m is also high compared to its major competitors.

An interview with the company revealed that Siemens KWU occupies the leading position in terms of provision of technology to the former USSR for large capacity gas turbines to be used in combined cycle systems, and participation in the installation of equipment using that technology in power stations in the former USSR.

Table 6.1 International Suppliers of Power Equipment (1994)

Company	Turnover	Boilers	Steam Turbines	Gas Turbines	Investment in Eastern Europe	Investment in Former USSR
ABB[1,2] (Asea Brown Boveri)	Group Total $28.3 bn (1993) Power Plants Division $7.9 bn (1993)	✓ (Combustion Engineering, USA)	✓	✓	✓ (Czech Republic [including PBS boilers and turbines, Brno], Poland, Hungary, Romania; 20,000 employees in Central and Eastern Europe; $465M income in Central and E. Europe, including £320M from Czech Republic and Poland)	✓ 16 companies employing 3,000 people including ABB Uniturbo [500 employees] and ABB Nevsky [full-scale manufacturing joint venture].
Siemens[3]	Group Total $49.5 bn (DM 81.6 bn) in 1993 Energy Division: $5.3 bn (DM 8.7 bn) in 1993	✗	✓	✓	✓ (Czech Republic, [Skoda Energy Division] $731M income [DM 1.2 bn] from Eastern Europe through 29 joint ventures[4])	✓ (Kalushsky Turbine Factory and Leningrad Metals Factory[5] [Interturbo])
GEC ALSTHOM[6]	Group Total: $14.06 bn (£9.41 bn) in 1993 Power Systems: $4.67 bn (£3.13 bn) in 1993	✓ (EVT [Germany]) (Stein [France])	✓	✓		(Kirov Works, St. Petersburg)

Company	Financials				
Mitsubishi Heavy Industries[7]	Group Total $24.3 bn (1993) Power Systems Products $6.1 bn (1993)	✓ (World's market leader in terms of installed capacity)	✓	✓	✓ (£43M sales in East Europe and CIS)	✗
GE Corporation (General Electric Company) 1993[8]	Group Total $60.56 bn in 1993 Power Systems $6.7bn in 1993	✗	✓ (20% of world market)	✓ (56% of world market)	✓ (Tungsram [Hungary])	✗ (but acquisition of 69% of Nuovo Pignone [Italy] which had received commitments for $1.6bn in pipeline equipment to Russia)
Westinghouse[9] 1993 Annual Report	Group Total $8.8bn in 1993 Power Systems £3.18bn in 1993	✗	✓	✓	✓ Transportation refrigeration manufacturing	✗
Deutsche Babcock[10]		✓	✗	✗		✓ (Bergemann-ZIO, Podol'sk, Moscow region Bergemann-Ilmarine, Tallinn)
Foster Wheeler Corporation[11]	$2.65bn in 1993		✗	✗		✓ (Joint venture with Hydrocarbon Technologies [HYTECH], Russia).

Table 6.1 continued

Company	Turnover	Boilers	Steam Turbines	Gas Turbines	Investment in Eastern Europe	Investment in Former USSR
Powell Duffryn[12]	$1.09bn (£728.5m) in 1994 $1.05bn (£700.4) in 1993	✓ Burners (Peabody and Hamworthy)	X	X	X	X
Babcock International[13]	Group Total: $1.12 bn (£0.75) in 1993 Energy and Manufacturing: $0.34 bn (£0.21 bn) in 1993	✓ (Babcock Energy)	X	X	X	X
Rolls Royce[14]	Group Total $6.25 bn (£3.56 bn) in 1992; Industrial Power Group $2.49 bn (£1.42 bn) in 1992	✓ (International Combustion Limited)	✓ (Parsons)	✓ (Parsons)	X	X (but supply of aero-engines to Tupolev for airworthiness approval of TU air frame)

Note:

✓ Denotes company activity in this area.

✕ Denotes no company activity in this area.

Sources for Table 6.1

1. ABB Annual Report and accounts - 1993.
2. *Financial Times*, 16th April 1993, p. 17 ('Pioneer Looks East for Profit'), *Fortune*, 2 May 1994, and ABB. Press Release, 27 June 1994.
3. Siemens annual Report and Accounts - 1993.
4. Financial Times, 27 November 1992, 'Czechoslovakia - Siemens Takes Control of Skoda Divisions'.
5. *Financial Times*, 18 August, p. 7 ('Siemens Arm in Turbine Pact with Russia').
6. GEC - Alsthom Annual Report and Accounts - 1993.
7. MHI Annual Report and Accounts - 1993.
8. General Electric Company, 1993 Annual Report, and 'GE in Europe', 1993.
9. Westinghouse 1993 Annual Report.
10. Deutsche Babcock, 1993 Company Profile and Organizational Structure.
11. Foster Wheeler Corporation 1993 Annnual Report.
12. Powell Duffryn, Annual Report 1994 and Catalogues from Hamworthy Engineering ('Hamworthy' and Combustion products for today's industries - Hamworthy Combustion Systems).
13. Babcock Annual Report and Accounts 1993.
14. Rolls Royce Annual Report 1992.

The gas turbine technology is being transferred to the largest civilian turbine manufacturer in the former USSR, and the only factory with large capacity power generation gas turbine manufacturing capability in Russia (*LMZ*). Turbines utilising this technology are to be installed in three power stations, namely the North Western (*Severno-Zapadnaya*) power station in St. Petersburg, the *Dzerzhinsk* Power Station in Nizhnyi Novgorod, and the *Mostovski* power station in Krasnodar.[3]

In all three cases, the installation will comprise a gas turbine, heat recovery steam generators, steam turbines and turbo-generators. These sets are to be supplied to greenfield sites in the cases of the St Petersburg and Krasnodar sites, and to an existing power station (block 3) at Nizhnyi Novgorod. The gas turbines are to be built by *LMZ* to Siemens designs, and the steam generators are to be built by the Ordzhonikidze Factory at Podol'sk to the designs of a German company. The steam turbines and turbo-generators are to be produced respectively by *LMZ* and *Elektrosila* (a major turbo-generator manufacturer in St. Petersburg) to their own designs, although Siemens are also providing some technical advice to *LMZ* on design modifications for improved efficiency of steam turbines. The overall project responsibility remains with the electricity generating utility which has commissioned the project (for example, *Lenenergo* in the case of the St Petersburg project) and the production of project documentation is delegated to a project- or scientific-research institute (for example, *VNI pri Energoprom*). *Lenenergo* place contracts on Russian suppliers based on that technical documentation, whilst supplies from foreign companies are handled by the foreign trade association responsible to the state-controlled joint stock company (*RAO-EES*) which co-ordinates electricity generation throughout the whole of the Russian national grid.

Siemens gas turbine technology transfer to *LMZ* has been achieved by the provision of design documentation related to final assembly, sub-assembly and component manufacture. The company has also located a group of engineers at the *LMZ* site who can explain the significance of the information contained in the technical documentation, the procedures to be followed in manufacture and assembly, and the stages and methods to be followed for approval to 'qualified supplier' status. These procedures are important for such critical components as turbine blades.

This technology is transferred to a joint venture ('Interturbo') which is jointly owned by Siemens and *LMZ*, with the Russian partner as the majority shareholder. Interturbo is housed in a separate building which belongs to *LMZ*, and is presently engaged mainly in assembly of the components delivered from the company's Berlin plant, transport of the finished turbine to the electric utility, and installation. It is intended that Interturbo will meet more of its own component requirements using *LMZ* facilities, until further machine tool investment has been made at the Interturbo site. It is also anticipated that Interturbo will finally produce some 90 per cent of its own requirements, using its own facilities and sourcing from outside suppliers.

The financing of the technology transfer requirements has required some special attention. The North West power station project also contains the Finnish electricity utility IVO as a partner, and a proportion of the electricity generated can be transmitted to Finland at an economic price, and some of the hard currency generated used to pay for the investment. In the cases of Nizhnyi Novgorod and Krasnodar a system of barter has been arranged to cover the equipment costs, and the Siemens organization contains a special department to arrange for barter trade. In addition, the cost of components supplied to *LMZ* can be partly covered by the re-export of turbines to third countries to raise foreign currency.

GEC ALSTHOM
Introduction GEC ALSTHOM is jointly owned by the General Electric Company plc and Alcatel Alsthom, and is divided into five divisions covering power generation, power transmission and distribution, industrial equipment, marine equipment, and transport. For the purposes of this research, attention will be paid to the activities of two divisions only, namely industrial equipment and power generation. The latter division designs and constructs turnkey power plants of all types and manufactures steam, gas and hydro turbines; electric generators and motors; boilers of all types; and environmental equipment. The division is also developing equipment for clean coal technologies.

The company had a turnover of some $9.13bn (7.928 bn ecu) in 1993/94, in which the power generation and industrial equipment divisions contributed some $4.06bn and $0.81bn respectively.[4] The company has sold gas turbines to Romania, and received orders for boilers based on fluidised bed technology in both Poland and the Czech republic. The company has also invested in a technical assistance agreement with the Kirov factory in St. Petersburg for the development and production of small gas turbines, and commercial arrangements with a factory in Ukraine. Close technical and commercial links have been established between EVT (Stuttgart) and Rafako (Poland) in the development of circulating fluidised beds, which could have applications in the former USSR, particularly in Ukraine. EVT is also engaged in a series of technical discussions with the Podol'sk and Belgorod boiler factories in Russia.

Nuclear Power GEC-ALSTHOM Engineering Systems Ltd (part of GEC ALSTHOM'S Industrial Equipment Division) considers that there are large market opportunities in the former USSR for the export of dry storage systems for spent nuclear fuel. The modular vault dry store (MVDS) consists of a concrete storage vault, permitting natural thermosyphon air coolant flow between sealed carbon steel fuel containers. The system has the advantage that the spent fuel can spend a shorter interval of time in wet storage in the 'coolant pond' following removal from the nuclear reactor, and the inherent advantage of dry storage over wet storage that contaminated resins are not generated during the storage process. The company gained licence approval for its generic design of systems from the United States Nuclear Regulatory Commission (US NRC) in 1988, and site specific design for the Fort St. Vrain (Colorado) MVDS which is now operational. The company has also gained experience in Eastern Europe through the Paks (Hungary) MVDS, which has received site specific licence approval and is now under construction. Proposals are currently being made for the use of MVDS technology for the storage of spent fuel from defence reactors in the USA.

The former USSR had some 47 nuclear power reactors including one reactor in Kazakhstan which is presently closed down, 29 in Russia of which twelve have been closed down, 2 in Lithuania and 15 in Ukraine of which two have been shut down. The Russian power stations are important in view of their quantity and generating capacity (19,843MWe), and also because they provide more than 12 per cent of the grid output in a time of reducing oil and coal production. Ukrainian nuclear power stations have a capacity of some 12,679 MWe, and produce some 32.9 per cent of the grid output, whilst the two Lithuanian power stations having a capacity of 2,370MWe produce some 77.9 per cent of the country's grid output (IAEA, 1994). The Lithuanian stations are both of the Soviet light water moderated gas-cooled design (RBMK), but of a later design than a similar unit which was molten down during the nuclear accident at Chernobyl. Furthermore, there are a number of RBMK reactors in Russia, although a few have been closed down.

There is clearly a demand for improved nuclear safety, and the company has experience in the storage of fuel from VVER 230 pressurised water reactors with the commissioning of the air-cooled store for this type of fuel from 4 reactors of this type at Paks. At present there is only some 800 MWe of VVER 230 in Russia, which could be decommissioned and replaced by improved generating capacity from the more modern VVER 13 stations.

The company's technology clearly has an area of application in the reduction of atmospheric pollution through the safe storage of nuclear fuels, and the possible further development of nuclear power to enable coal-burning and associated pollution to be reduced.

Gas Turbines and Compressors GEC ALSTHOM has extensive experience in the supply of power engineering equipment to the former USSR, as individual firms which are now members of the group have jointly supplied some 200 turbine compressor drive units for gas pipelines. The company has also built on its previous experience of supply to the former USSR, to develop Russian manufacture of gas turbine equipment for power generation and mechanical drive. This has been achieved through the signing of a technical assistance agreement between the European Gas Turbine Company NV (EGT), a subsidiary of GEC ALSTHOM and General Electric of the US (GEC ALSTHOM, 1991) and the Kirov Factory in St. Petersburg. The Kirov Factory will manufacture and process various parts for the turbines in Russia, whilst EGT will supply the blading, rotors, nozzles and combustion chambers to enable complete assembly and testing to be carried out in St. Petersburg.

Initially, the agreement is confined to 25MW turbines and a progressive transfer of technology is envisaged based on 30 twin shaft compressor units with the possibility of power generation units later. Entry into power generation would enable Kirov to meet small combined heat and power plant demands in the former USSR. The technology transfer has required no product research and development by Russian organizations, although attention has been required to process planning in order that the requisite tolerances be achieved. These tolerances have to be met using existing facilities within the factory, and modifications have consequently been required to previously defined process sequences to reflect the realities of machinery and tooling already in use. Three 25MW units have been produced to date for installation in gas pumping lines, and it is envisaged that production will be increased to an output volume of ten units per year.

The transfer of this technology will have an impact on reduced environmental pollution through access to the option of dry low NOx burners which can be used on the 25MW turbines, and retrofitting of old units should enable a figure of 50ppm ($100mg/Nm^3$) to be achieved. The company is also of the view that atmospheric pollution problems could be further reduced through improvement in the effectiveness of the electricity transmission system, using improved designs of switchgear and cabling. In addition, the use of alternative fuels and the development of alternative clean coal technologies such as integrated gasification combined cycle and pressurised fluidised beds, may also reduce the overall levels of atmospheric pollution.

Since GEC ALSTHOM, through European Gas Turbines NV, has already established a mechanism for the transfer of technology for the manufacture of gas turbines, it is clearly interested in extending its power plant activities in Russia. The company is cautious about part-ownership and operation of a power station, however, because of questions over the commercial viability of power stations in Russia in the current climate of cash flow crisis.

Any initial investments in Russia are likely to be financed by countertrade back into Western countries, although investments may also be extended into the manufacture of products and components for which a Russian location has a cost advantage and there is a technical capability to manufacture the product. To a very great extent, however, the development of future business in Russia is contingent upon economic stability, particularly in terms of currency convertibility and inflation. In addition, many of the property rights and obligations of investors require further clarification, and concern over the security of receipt of royalty payments also needs to be addressed. This may be achieved through the earmarking of tranches of multilateral funding to be reserved for the payment of royalties.

Rolls Royce Industrial Power Group

Rolls Royce Industrial Power Group designs and manufactures a range of power engineering products and services. These include steam generators; steam turbo-generators; aero-derivative gas turbines and nuclear engineering products. The Industrial Power Group achieved a turnover of some $2.5bn in 1992, within a total turnover of some $6.25bn for the parent company's total turnover from its aerospace and power engineering products.

International Combustion Limited (ICL) (part of the Group's power engineering division) produces a wide range of boilers, combustion systems and pressure parts, and now includes steam generators for combined cycle systems within its product profile. The company has achieved particular market success in the design and manufacture of low NOx burners for both wall-fired and tangentially-fired boilers. These combustion systems have been sold successfully in the UK and US markets in view of their capability to meet EU and US federal regulations for NOx emissions and the company has also achieved a significant share of the Far Eastern market.

In view of the size of the UK market, and the pace of growth in the Far East, it is apparent that these regions currently present far more attractive market possibilities than the former USSR. International Combustion has responded to a series of requests from both Eastern Europe and Ukraine, including protracted discussions with the Rafako boiler plant in Poland, but no definite delivery contracts or technology transfer arrangements have yet been agreed in that region. The company is aware of the future sales potential in the former Soviet region, however, and also of the capability of the region to product materials and labour intensive fabrications in a cost-effective fashion, which may be used to service other markets. At the time of writing, however, International Combustion has been made available for sale (*Financial Times,* 1997b), and the position of this company in the Russian market is therefore less certain than when this case research was conducted.

Rolls Royce Industrial and Marine Gas Turbines Limited has previously sold a number of gas turbines into the former USSR for the driving of gas pipeline pumping stations, based on the Avon aero-engine. There are consequently market opportunities for the delivery of spares for these engines, and also for the delivery of the newer generation of aero-derivative industrial gas turbines based on the RB 211 engine. These power units could be used in gas pumping installations as previously, or also as 40-50 MW drive units for use in power generation including combined cycle installations. Potential may exist for the manufacture of components required for engine spares in the former USSR, and these opportunities could also be linked to possible future sales of RB 211 engines from the parent company's aerospace group to power the TU204 aircraft (Hill and Hay, 1993, pp. 43-6). There are also opportunities for the transfer of the group's gas turbine technology to industrial applications, including the conversion of military jet engines to major units in industrial gas turbine configurations. A joint venture has recently been established with the Moscow-based Lyulka-Saturn company for the purpose of adapting the Sukhoi 27's jet engine to power generation (*Financial Times,* 1997a).

An important feature of the group's power engineering division's recent experience, however, has been the establishment of a strategic alliance with the American Westinghouse corporation, in which aerospace technology was used to improve the performance of new marine gas turbines to be installed in US Navy vessels. This strategic alliance with Westinghouse presents further opportunities for power generation business, as larger capacity Westinghouse gas turbines (or those produced by other members of the alliance, that is, Fiat and Mitsubishi Heavy Industries) can be fitted alongside C.A. Parsons steam turbo-generators in combined cycle systems, and sold into those markets where Parsons have a competitive edge. In addition, the technology of the large gas turbine drive units may also be further improved through the transfer of aerospace technology, particularly with regard to blade and vane design, and high temperature combustion conditions. The possibilities consequently exist for the further development of combined cycle power plant technology transfer in the former USSR, with a view towards the utilisation of manufacturing facilities within that region. At the time of writing, however, Parsons Power Generation Systems is in the process of being acquired by Siemens KWU (*Financial Times*, 1997b), and the effects of this takeover on both companies' activities in the Russian market are therefore awaited.

Babcock Energy
Babcock Energy is a long-established designer and manufacturer of power engineering equipment, particularly fossil-fuelled boilers and nuclear powered steam generators for marine and power engineering applications. In recent years, the company has developed its product range to include lower pollution combustion equipment such as low NOx burners, and heat recovery steam generators, and has purchased licences from Babcock-Hitachi for the manufacture of flue gas desulphurization equipment. The company was previously part of the British subsidiary of the American-based Babcock and Wilcox Corporation but following takeover and subsequent de-merger, the company emerged as part of Babcock International, with a turnover of some £252 million.

Babcock Energy has had an interest in the markets of the former Soviet Union since the early 1980s. Presentations have been made by senior technical and sales delegations to leading members of the All-Union (now All-Russian) Thermal Energy Institute (VTI) in Moscow, and visits have also been made to boiler making factories at Podol'sk and Taganrog, the Polzunov Central Boiler and Turbine Institute (*TsKTI*) at St. Petersburg, and several power stations in various republics of the former USSR. These visits have provided sufficient information to demonstrate a high level of competence in boiler making in the former USSR, and an understanding of the environmental problems to be tackled in the field of electric power generation.

Furthermore, factories in the former USSR have sufficient expertise in the manufacture of finned tubing and general heat exchanger construction to manufacture heat recovery steam generators for use in combined cycle systems. The fin-making technology is not necessarily as advanced as that used in Western factories, but the heat exchangers should be capable of operating at efficiency levels of only some 2 per cent lower than those in Western countries.

The company has been mainly engaged in proposals for technology transfer for low NOx burners operating with Ekibastuz coal, and experimental work is currently being carried out in a power station located in the eastern Urals *(Reftinskaya GRES)* to arrive at nozzle design parameters and power station operation procedures for this type of coal. Furthermore, there may be scope for assistance in the resolution of problems of metal creep in boilers operating at supercritical conditions.

The assimilation of Western technology to resolve the problems of atmospheric polution, however, is contingent upon a reconsideration of some of the design procedures used in the former USSR, to evaluate whether some of these procedures may have been over-prescriptive and inflexible. If this proves to be the case, modified codes of design, manufacturing and operating practice may require development. The major barriers to technology transfer, however, will probably be commercial, in view of the present shortage of funds for investment purposes, and the lack of economic incentives for power station managers to adopt over-firing procedures. In addition, it is far from clear as to what products can be offered to Western companies as counter-trade for the purchase of the requisite technology, and the savings·to be gained from lower Russian production costs to provide an alternative source of supply have yet to be fully demonstrated.

In spite of these commercial barriers, however, the company is of the view that sufficient technical competence and manufacturing expertise already exist in the former USSR as technical pre-requisites for transfer of the appropriate technologies. Some of the commercial problems could be reduced through continued provision of multilateral aid from either European or UK government sources. The company's future activities in the former USSR await to be seen, however, as it has recently been acquired by the Mitsui Engineering and Ship building of Japan *(Financial Times*, 1995).

Deutsche Babcock AG
Deutsche Babcock AG is a well established power engineering company with divisions engaged in boiler-making, pipework and construction projects related to electrical power generation. The company has recently established a joint venture between its Bakke-Durr AG (BDAG) subsidiary, and the Ordzhonikidze Engineering Factory in Podol'sk for the construction of cooling towers and steam condensation equipment. BDAG holds 49 per cent of the shareholding in the joint venture, with 46 per cent held by the

Ordzhonikidze Factory. The remaining 5 per cent is divided between the foreign construction division of the Russian stockholding company for international investment in power station construction *(RAO-Zarubezh Energostroi)* and a Krefeld based organization for trade with Russia (Gesselschaft für Russische und Internationale Wirtschaftskontakte). In addition to this joint venture, Deutsche-Babcock is also providing technical assistance to the Podol'sk factory in the general field of boilermaking production technology.

Ahlstrom Boilers
Ahlstrom Boilers is part of Ahlstrom Pyropower - the division of the Finnish-based Ahlstrom group which specialises in advanced combustion technology for utilities and industry. The division's headquarters are in San Diego, California, but the company also has business units operating in Kobe and Tokyo; and in Varkaus, Helsinki and Kaarina. Boiler making facilities are located in Varkaus (Finland) and Sosnowiec in Poland.

The company has developed a wide range of advanced combustion technologies, including oil and gas-fired boilers, waste-heat boilers for industrial processes, and heat recovery steam generators for use in combined cycle and co-generation systems. Of particular interest to this project, however, is the company's expertise in the development of Pyroflow circulating fluidised bed boilers, which can capture up to 95 per cent of sulphur dioxides through limestone fed into the combustion chamber, and the use of staged combustion to reduce emissions of NOx. Nitrogen oxides can be further reduced by injecting ammonia, eliminating the need for expensive SCR systems.

The systems can use a wide variety of fuels to produce outputs from the boiler between 10 and 60MW, and electricity outputs varying between 3 and 20MW. In addition, Pyroflow boilers have been installed in progressively larger plants, with scale-up based on over 400 years of total operating experience. One plant in operation in Nova Scotia, for example, has an output of some $420MW_{th}$ ($165MW_e$ net). The company has also installed some 75 bubbling bed boilers from 10 to $150MW_{th}$, and development work is in progress on pressurised circulating fluidised bed systems and integrated gasification combined cycle systems.

It is apparent therefore that the company has a wide range of advanced combustion technologies available for transfer to the former USSR, and the existence of a trading base in Finland to deliver this technology. It is apparent that there is a large potential for this type of equipment in the former USSR, particularly in those regions containing coal deposits but limited supplies of natural gas (for example, Ukraine). The Russian reaction to proposals in the power generation sector has been slow, however, as a result of a shortage of funding.

The reaction to proposals in the non-ferrous metals sector has been more hopeful, partly because of the possibilities of raising funding for the projects through the sale of non-ferrous metals on the international market. Proposals have been made for the fitting of waste heat recovery steam generators to be fitted above smelters, to enable the waste heat to be used for steam generation, and the gases to be cleaned as they leave the filter. The potential use for these systems is very wide in the non-ferrous smelting factories of Russia, Ukraine and Kazakhstan.

US Power Engineering Companies
Two large American companies (GE Corporation and Westinghouse) have expertise in turbine design and nuclear engineering, but have no large boiler making facilities. GE Corporation with a turnover of $6.7bn from its power systems division is a major turbine manufacturer, and possibly the world's largest manufacturer of gas and steam turbines. Furthermore, the company has an established sales office in the former USSR and has established strategic alliances with several other power engineering companies, including European Gas Turbines (EGT) - a subsidiary of GEC ALSTHOM. GE has embarked on joint venture arrangements between its power systems division and the Kirov Factory in St. Petersburg for the production of gas turbines (GE, 1995).

Westinghouse is also a large manufacturer of steam and gas turbines, with a demonstrated capability to install these machines in a combined cycle configuration. In recent years, Westinghouse has entered into strategic alliances with power engineers having established positions in other markets (that is, Rolls Royce Industrial Power Group in UK and Mitsubishi Heavy Industries in Japan and the Far East). At the time of writing, however, it appears to be less involved than either GE Power Systems or its German competitors, in the conclusion of technology transfer and joint venture arrangements with Russian factories.

Foster Wheeler is a large American boiler manufacturer, which also manufactures a range of low NOx burners, but there is little published evidence of a high level of involvement in this field in the former USSR by this company, although the company has established a joint venture in the general field of hydrocarbon technologies (see Table 6.1).

Japanese Power Engineering Companies

Mitsubishi Heavy Industries (MHI) has grown very rapidly in the boiler, steam and gas turbine and nuclear engineering markets, and is probably now the world's market leader in terms of volume of boiler production because of the growth in power generation requirements in the Pacific Rim. It has also purchased the know-how for several combustion technologies from the CE boiler subsidiary of ABB, and has the right to develop some of those technologies. From data shown in Table 6.1, it is quite probable that its volume of output of power engineering equipment exceeds that of Siemens and possibly ABB, although it is difficult to compare volumes of output of these companies because of the different products included in each company's power engineering product range. The company also has a high level of expertise in the design and manufacture of pollution control equipment, partly driven by the Japanese government's stringent requirements for pollution reduction, and has developed a simplified semi-dry flue gas desulphurization system which is initially being exported to China. Although MHI's traditional markets have been Asia, America and the Middle East, there would appear to be no reason why it should not attempt to penetrate the Russian market, as its parent company has an office in Moscow. Furthermore, the company could follow the traditional path of many Japanese firms and invest in facilities in Siberia (Manezhev, 1993).

In addition, the Hitachi group is a leading company in the area of design and production of equipment for reduced atmospheric pollution, and reference has previously been made in this chapter to the licensing of this equipment by the Babcock Energy Group.

Energy Efficiency and Conservation Companies

The previous parts of this section of the chapter have described the business activities of Western power generation and power engineering companies in the former USSR, including the implementation of procedures and the installation of plant for increased power generation efficiency. Atmospheric pollution can also be reduced, however, by the conservation of energy by major users of electrical power, as the savings thereby gained can reduce demand for electricity with consequent savings in fuel burnt and subsequent pollution. The following description provides information on a UK company engaged in this type of activity in the former USSR.

LINDEN Consulting Partnership

The LINDEN Consulting Partnership comprises some four partners having a combined experience of some ninety man years in energy conservation. This experience has been gained in the UK on projects for a number of private companies and for the Energy Technology Support Unit, and on a number of international projects funded by the European Commission, the United Nations Development Programme, and the World Bank. The partnership has also carried out a number of projects in the former USSR, including four projects in the Russian Federation, one in Uzbekistan and one in Moldova. One of the projects in the Russian Federation was supported through the UK Know-How Fund, and included energy audits and the training of Russian engineers in the area of energy conservation. This training took place both in Russia at the Moscow Power Engineering Institute, and also in the UK at a number of organizations including drop-forging factories and power generation companies.

The three other projects in the Russian Federation were all funded through the TACIS Programme. The first project was carried out in 1991 and paid particular attention to energy conservation in the food industry. Subsequent investigations were carried out in Novosibirsk and Ekaterinburg in 1993, and Rostov and Pyatigorsk in 1994, to identify and define energy conservation projects. As a consequence of these investigations, it is intended to set up a series of six demonstration zones which will include both industrial and domestic examples, with particular attention being paid to the steel industry.

RUSSIAN POWER ENGINEERING SUPPLIERS

The Ministry of Power Engineering was formerly responsible for the design, development and production of power engineering combustion and turbine equipment across the former USSR. Much of its product profile was for civilian application rather than military use, as most of the military production ministries would have had the necessary facilities to manufacture the specialised power engineering equipment required for naval, aeronautical and armoured vehicle applications. Furthermore, the Ministry's product profile would probably also have included nuclear engineering, to supplement engineering capacity in the Ministry of the Atomic Energy Industry (*Minatomenergoprom*). The factories formerly responsible to *Minenergomash* engaged in the manufacture of power engineering products relevant to this current research include boiler factories at Taganrog, Podol'sk (near Moscow), Barnaul, Byisk, and Belgorod, and major gas and steam turbine factories in St. Petersburg (Leningrad Metals Factory [*LMZ*] and the Nevskyi Engineering Factory), and Khar'kov in Ukraine (variously described as the *Kirov* Turbo-Generator Factory, the Khar'kov Turbine Factory [*KhTZ*] or *Turboatom*). Factories engaged in mainly small and medium-sized turbine production are located in Kaluga and in Ekaterinburg (formerly Sverdlovsk) (that is, the Urals Heavy Engineering Factory [*UZTM*]). Large power engineering factories have also been established in Volgodonsk (*Atommash*) and at other locations in the St. Petersburg region (Izhorsk). *UZTM* was responsible to the Ministry of Heavy Engineering (*Mintyazhmash*), although it is not entirely clear from published sources whether the Izhorsk and *Atommash* factories with their nuclear engineering product profiles were responsible to *Minenergomash,* or the defence ministry *Minatomenergoprom* (Rosengaus, 1986; Rosengaus, 1987, pp. 31-51). Research and development facilities were provided by a number of additional establishments, but particularly by the Central Boiler and Turbine Institute (sometimes known as the Polzunov Institute) in St. Petersburg. This institute achieved international recognition for research and development in the field of sub-critical steam operating conditions, and produced many of the early designs for the Taganrog Boiler Factory. Research on heat transfer in nuclear energy boilers has been carried out by the Institute of Nuclear and Boiler Engineering (*Atomkotelmash*) at Rostov-on-Don (Cooper, 1988).

In addition to the range of 'civilian sector' turbine factories referred to in the previous paragraph, turbo-machinery was also produced in factories responsible to ministries in the defence sector including aero-engine factories at Nikolaev, Rybinsk, Samara, Tula and Ufa, and the *Kirov* general engineering factory in St Petersburg. The main product lines of these factories have been gas turbines for use in aviation and marine propulsion, although power engineering variants have also been produced for gas pumping along pipelines, and new models of power generation turbines are also being developed (see Chapter 4).

Following the fragmentation of the USSR into independent states in 1991, subsequent reduction in status of former Soviet industrial ministries to Russian federal departments for the administration of industrial facilities on Russian territory, and the introduction of legislation for privatization of many factories, it is difficult to assess the current status of many of the power engineering factories previously responsible to *Minenergomash* and other ministries. The majority of power engineering factories and research and development establishments in the St. Petersburg district formed the

Energomash Association in that city, but subsequently every civilian power engineering factory became privatized and some factories transferred to membership of the *Energomash* joint stock company. Many of the larger St. Petersburg factories subsequently decided to create an alternative company, however, named as the *Energomashinostroitel'naya* Corporation. The ownership status of the Russian aero-engine factories will still be influenced by the Ministry of Aircraft Production in view of the importance of their output to Russian defence strategy, although the Nikolaev factory is now located in a separate independent state (Ukraine).

The following parts of this section of the chapter provide information on three factories in the St. Petersburg region selected for case research in view of their capabilities in the fields of gas turbine and combined cycle technologies, and the use of clean coal combustion. The information contained in these descriptive accounts was obtained from visits to these establishments in the summer of 1994 and the spring of 1995 and 1996, for interviews with senior executives.

Leningrad Metals Factory (*Leningradskii Metallicheskii Zavod* [*LMZ*])

The Leningrad Metals Factory (*LMZ*) is the largest manufacturer of turbines for power generation applications in the former USSR. The factory was established in 1857 as the St. Petersburg Metals Factory and commenced the manufacture of turbines for the Soviet electrification programme in 1924. Since that date, *LMZ* turbines have been installed in 70 per cent of the power stations in the former USSR, and in twenty eight countries worldwide.

The company's product range includes steam, hydro- and gas turbines, but the majority of the output has consisted of steam turbines. Some 65 per cent of generating capacity in thermal and nuclear power stations in the former USSR, and 9 per cent worldwide, is provided by *LMZ* steam turbines having capacities ranging from 60/80MW for subcritical machines used in combined heat and power stations, to 1400MW machines operating in supercritical conditions. The technical parameters of these machines are considered to be competitive with those of Western manufacture, although there is scope for improvements in the systems of automatic control. These improvements can be partly achieved by the purchase of Western automation equipment. The company is also an established manufacturer of hydro-turbines from 4MW to 700MW capacity. The manufacture of hydro-turbines has accounted for some 25 per cent of the factory's output, and machines have been delivered to all of the states of the former USSR, China, Canada, Mexico and South America, and Sweden.

LMZ is also a manufacturer of gas turbines for use in load peaking and semi-peaking conditions, although the output of these machines has accounted for less than 10 per cent of the company's output. The capacity range varies from 10 to 150MW capacity, working at inlet temperatures of some 950°C. This range is currently being upgraded, with particular attention being paid to the design of rotors, blading and the combustion chamber, and inlet temperatures have now been raised to 1150°C.

The market opportunities for the company can consequently be considered to fall into three main groups, namely the continued supply of steam and hydro turbines and associated spares to the Russian and international markets, particularly China; and the development of gas turbines for the Russian market, with future export market possibilities. The scope for success in the market for new steam and hydroturbines, however, will be influenced by funding for the renovation of conventional thermal and hydro power stations in the former USSR, and the opening up to Western competition of former captive markets in the former USSR and Eastern Europe. In view of these uncertainties over competition and funding for new power stations, it would appear that a major short to medium term opportunity for the company is the provision of spares. The market for hydro turbines may be more promising than for steam turbines, however, as there is still scope for the further exploitation of water resources in some of the former republics of the former USSR, and hydro power stations have the advantage of almost zero atmospheric emissions.

Response to opportunities for gas turbines would appear to need some further product development, however, and this is being achieved in two ways, namely the updating of the company's present range of 100MW (GT100) and 150MW (GTE150) range of turbines being replaced by the GT-140; the licensed production of 150MW and 60MW turbines to Siemens' designs; and the licensed production of 40-60MW capacity to the designs of the Nevsky Factory. A market is considered to exist for these turbines in combined cycle configurations throughout the former USSR, in view of the improved efficiencies available from these systems. 150MW machines to Siemens designs are currently being assembled for delivery to the *Severno-Zapadnaya* power station in the city of St. Petersburg together with the delivery of *LMZ* steam turbines, and other opportunities for this gas-fuelled machinery may also exist in other regions of the Russian Federation. *LMZ* is supplying between 7 per cent and 10 per cent of the components of the first turbine, but it is anticipated that the share provided by *LMZ* will gradually increase for subsequent machines to a level of 100 per cent of the components for the fifteenth turbine.

The joint venture with Siemens has consequently quickly provided an extension of the product range, whilst the factory's technical facilities can continue to design and develop an updated replacement for the GT100 and GTE150 machines. The company has previously shown its capability to meet the requirements of such Western agencies as Lloyds and TUV through its volume of exports to Western markets having stringent quality requirements. The co-operation with Siemens, however, has now provided an impetus for the factory to extend its quality management expertise beyond that of quality control and traceability of products and components, to the workplace control of process conformance which is widespread in Western factories.

The future development of the factory, therefore, is dependent upon a range of factors, the most important being the uncertainty over market conditions for new turbines within the Russian Federation. The company has almost a monopolistic position in the Russian Federation with regard to the supply of gas, steam and hydroturbines for power generation, with the only competition for gas turbines coming from aero-engine manufacturers (for example, at Rybinsk, Ufa and Samara) and for steam turbines from the Urals Heavy Engineering Factory at Ekaterinburg. Previous competition from the Nikolaev Turbine Factory and the Khar'kov Turbine Factory has virtually disappeared since 1992, as both Ukrainian factories are now regarded as foreign suppliers to Russia.

The other important factor, however, is the strategic direction provided to the factory under the new ownership structure. At present 17 per cent of the shares are owned by a finance and trading company (Mardima), 10 per cent by Siemens (purchased in hard currency), and 73 per cent by the *Energomashinostroitel'naya* Corporation. These latter shares were acquired by the purchase of vouchers from the labour force, and is a pattern which has been repeated in most other turbine producers in the former USSR. The development of *LMZ* is therefore influenced to a great extent by the strategy of the *Energomashinostroitel'naya* Corporation. *LMZ* is also attempting to preserve assets and guarantee its source of supply in the Leningrad region by the cross-purchasing of shares in its supplying factories in that region. It remains to be seen whether this is a more competitive stance than supplier development outside the region, but many alternative sources of supply are in their turn owned by large corporations in the former defence sector.

The Leningrad Turbine Blade Factory (*Leningradskii Zavod Turbinnykh Lopatok* [*LZTL*])

The Leningrad Turbine Blade Factory (*LZTL*) was established in 1964 and became a joint stock company in 1992. It occupies a site of some $276,500m^2$, and employs approximately 2,500 people. Forty four per cent of the companies shares are owned by the *Energomashinostroitel'naya* Corporation or its affiliates (19 per cent by the *Energomashinostroitel'naya* Corporation, 5 per cent by an *Energomashinostroitel'naya* subsidiary, and 20 per cent by *LMZ*), and 33 per cent by the Sturmhamer Corporation (Amsterdam). The remaining shares are owned in small blocks by small companies, or by the factory's employees.

The product range of *LZTL* includes finished turbine blades, the provision of stampings, castings and forgings for subsequent machining, and rolled metal sections. It has a monopoly position in the supply of cylindrical blade stampings for medium and low pressure steam turbines and stationary gas power turbines including those used for gas pumping units of 10 and 25MW, and the company also holds 30 per cent of the CIS market for all types of finish machined blades. The factory is equipped with a range of up-to-date equipment in the forge, foundry and machine shop including directional control casting machines and multi-axis numerically controlled machine tools; and has recently received orders from ABB for nine types of turbine blade, and has also produced an experimental batch for Westinghouse. The quality management and inspection procedures were carried out according to an agreed plan and documentation with these customers, although the company would also like to receive either Lloyds or TUV certification. As different customers use different certificating authorities, however, it is difficult for the company to know which is the best authority to choose in view of the initial investment required to secure certification. At present, a quantity of blades is being produced for a marine engineering customer, and the costs of the requisite certification will probably be shared.

The major customer, still remains as *LMZ*, and other turbine manufacturers such as the *Turboatom* factory at Khar'kov and the Urals *Turboatom* Factory at Ekaterinburg. In view of there being no new power stations being built at present, the company is relying on spares for renovated turbines as its main means of securing business. It will also need to invest in new machinery for the production of cast cooled blades, in order to complete in international markets, but also to compete in the supply of blades for the GTE-150 and GTE-115 being produced by domestic companies (*LMZ* and Turboatom).

Such investment will also be required to become established as a source of supply of blading for the Siemens 94.3 turbine being made under licence by *LMZ*. Furthermore, some more attention will be required to be paid to quality management systems if supplies to the West are to be increased. Nevertheless, in spite of these necessary required improvements, the company can be viewed as a proven supplier of turbine blading and a potential supplier of blading to the more advanced specifications required by international markets.

The Central Boiler and Turbine Institute (*Tsentral'nyi Kotelno-Turbinnyi Institut [TsKTI]*), St. Petersburg

The Central Boiler and Turbine Institute was established in the 1930s as a combustion development establishment for both boilers and turbines, and it subsequently emerged as a leading research and technical organisation in the power engineering industry alongside the All-union Thermal Energy Institute in Moscow (*VTI*). Its importance grew with the subsequent expansion of the Soviet electrification programme, providing research and development expertise to boiler, turbine and other power engineering industries, whereas *VTI* was a constituent organisation of the electricity power generation industry. The work of these two establishments frequently overlapped, however.

The range of scientific and technical publications from scientists and engineers within *TsKTI* is impressive in terms of its breadth and depth, and the institute was a pioneer in combined cycle technology in the former USSR. Systems to the institute's design were installed in the Nevinomysskaya state power station (170MW in 1972), and the Moldavian state power station in 1974 (2 × 250MW) using Khar'kov 35MW gas turbines, and either Khar'kov or *LMZ* steam turbines and boilers from Taganrog. Three systems were also subsequently installed including a 25MW unit in the *Krasnyi Treugol'nik* factory in St. Petersburg, but there have been no further installations in the last sixteen years. To a certain extent, the lack of fully suitable domestically produced gas turbines acted as a barrier to further development.

The institute has subsequently developed two proposals for combined cycle systems, including a 40MW system with a steam recovery unit in Vologda (using a 25MW Nevsky Zavod gas turbine, 15MW Kaluga steam turbine and a Belgorod boiler) and a 250MW integrated gasification combined cycle system at the Kirov TETs (CHP) No.5 (using a 115MW Khar'kov gas turbine, an *LMZ* steam turbine, and a boiler from Taganrog). These proposals are presently awaiting financing even though combined cycle systems have now been designated as the favoured technical option for new gas fired power stations, alongside a programme of 'topping-up' using gas turbines in existing conventional gas-fired power stations.

The institute has also developed two working proposals for circulating fluidised bed systems, one with a boiler capacity of 160 tons per hour and another at 500 tons per hour, with the Barnaul Boiler Factory having the capacity to commence production of these units when the orders are finalised. At present, however, 140 million roubles is owed by customers to power stations, and there is therefore little available funding for new projects. This is leading to the situation of chronic underfunding within *TsKTI*, with no salaries having been paid for the last eighteen months. Nevertheless there is a high level of technical expertise in boilermaking and combustion engineering within the institute which could be readily used by a Western company. Furthermore, the institute has extensive contacts with boiler-making factories within Russia with associated access to manufacturing capacity. At present, however, it appears that Western manufacturers appear to prefer to work with the boilermaking factories directly, rather than through the medium of *TsKTI*, although the institute has recently signed a co-operation agreement with Steinmuller (Germany) for the transfer of technology for heat recovery units, and this agreement also includes the Taganrog Boiler Factory.

CONCLUSIONS

This chapter has identified the major Western participants likely to be engaged in the transfer of technology related to the reduction of atmospheric emissions in the former USSR, and described their activities. These activities are summarised in Table 6.2, and divided into those companies for which information was obtained mainly by case research, and those for which information was obtained from publications and press reports. This table highlights the high level of joint venture activity (for example Siemens, ABB and GEC ALSTHOM) in gas turbine technology which has previously been identified as a technology of growing importance in view of the anticipated continued use of gas as a major fuel (see Chapters 2 and 5) and the high efficiencies available from combined cycle systems (see Chapter 4). The high level of joint venture activity in gas turbines, however, will also be influenced by the interest of Western firms in the gas transportation market. Table 6.2 and the case research also reveal interests from a number of companies in the area of Low NOx coal combustion technology.

The information presented in this chapter has also illustrated that there is a well-established infrastructure of power engineering factories and research and development facilities in the former USSR, although the majority of these facilities are in the Russian Federation which may have some effects in the other former Soviet republics. There is also evidence in the case studies of a capability for these factories to assimilate Western technology and to achieve Western standards, although more detailed research is required on the capabilities of particular key processes before definite conclusions can be reached. Such process capability information is proprietary information, however, and proved impossible to obtain in this current research; and no published information could be found on typical tolerances and process specifications for the power engineering industry.

It can therefore be concluded that as well as the existence of relevant Western technologies for the reduction of SOx and NOx emissions from power stations as described in Chapter 4, there is also evidence of a technological and manufacturing capability in the former USSR to facilitate the assimilation of these technologies. It is also apparent that Russian power engineering factories are enthusiastic to establish closer contacts with Western power engineering companies as evidenced by the following reports of co-operation agreements additional to the cases of *LMZ* and Siemens, and *TsKTI*, Taganrog Boiler Factory and Steinmuller, described in the previous section of this chapter. The reported agreements are:

1. Joint venture ZiO (Podol'sk)/Cockerell for heat recovery units for combined cycle systems.
2. Narofominsk 'Energodetal' (AO 'Gasenergoservis')/Siemens and Thyssen for turbine blades;
3. Ryazan' and Gusinoozersk GRES/Lurgi, Lentjes and Bishopf for sulphur cleaning equipment;
4. Severnoi TETs (Moscow)/Topsoe Denmark for NOx cleaning equipment (*Energetik,* 1994).

It is also apparent, however, that Russian power engineering factories and research establishments have undergone major changes as demand has rapidly fallen in the former USSR and the previous socialist countries of Eastern Europe, and the process of privatization has also raised questions about long term viability. Furthermore, it appears that the basis of competition in the power engineering industry may lead to rivalry between the former industrial ministries in the civilian and defence sectors respectively, with production allocated to their now-privatized subsidiaries. In spite of these uncertainties, however, it is evident from the case studies and list above that a number of Western companies are still willing to engage in joint technical and production operations with Russian factories and research and development establishments, in the anticipation of increased business opportunities in this region.

The case studies also reveal a positive Western view of the capability of Russian factories to assimilate the relevant technologies, revealing that many of the technical preconditions are in place, and that many of the organizational barriers to transfer are being overcome through joint venture arrangements.

Furthermore, the assimilation of Western process engineering expertise is being facilitated by the location of Western teams in Russian factories, for training of local personnel in process conformance or the adaption of Western process sequences to suit machines and tooling at the recipient's site. In addition, attention is being paid to supplier selection to meet advanced standards and quality consistency; although the selected suppliers may not always match those previously used by the Russian end-user if the above-cited report is correct of Siemens' co-operation in *'Energodetal'*, whereas *LMZ* usually source from *LZTL*.

The barriers to technology transfer are mainly identified as commercial and financial rather than technical, however, suggesting increasing investment in those projects where foreign currency earnings are possible (for example, power stations near to the Finnish border), projects part-financed by Western governments or multilateral agencies, projects part-financed by materials sales to the West, some of which may be payments-in-kind from industrial customers. The case study research described in this chapter has shown that the most successsful companies engaged in the transfer of power engineering technologies to the former-Soviet market, have highly developed levels of expertise in counter-trading activities. The ability to counter-trade was always a factor influencing success in exporting to the previously centrally planned Soviet economy, but it has now become even more important in the present conditions of transition and associated shortages of domestic investment capital in the region. These issues are explored further in Chapter 7 below.

Table 6.2 Areas of Mutual Interest

ORGANIZATIONS IN RUSSIA	FUEL PROVIDERS	BOILER R&D ESTABLISHMENTS	BOILER FACTORIES	GAS TURBINE FACTORIES	STEAM TURBINE FACTORIES	UTILITIES
WESTERN ORGANISATIONS						
(a) Information based on published reports and case research*						
Coal Research Establishment	✓	✓	✓	X	X	X
ABB	✓	✓	✓	✓✓	✓	✓
Siemens	X	✓	✓	✓✓	✓	✓
GEC ALTSHOM	X	✓	✓	✓✓	✓	✓
IVO	✓	✓	✓	✓	✓	✓
Babcock Energy		✓	✓	X	X	X
Rolls Royce IPG	✓	✓	✓	✓	✓	✓
(b) Information based on published and company reports only						
MHI	X	✓	✓	✓	✓	✓
GE**	X	X	X	✓✓	✓	✓
Westinghouse	✓	X	X	✓	✓	✓
Deutsche Babcock**	✓	✓	✓✓	X	X	✓
Foster Wheeler	✓	✓	✓	X	X	✓
Powell Duffryn	✓	✓	✓	X	X	✓
Powergen , National Power	✓	✓	✓	✓	✓	✓

Notes

✓	Denotes likely mutual interests
✓✓	Denotes established joint venture
X	Denotes unlikely mutual interests
*	Case research material was obtained from these companies by means of company visits and discussions with senior executives
**	Additional material was obtained from discussions with senior executives at GE and Deutsche Babcock.

From the perspective of business developments for UK companies the research has demonstrated that UK-owned power engineering companies are generally not so well established as their West European and American-owned competitors in the markets of the former USSR for technologies for gas turbines, heat recovery steam generators, and advanced coal combustion. Furthermore, the market will probably be even more competitive if Japanese companies become more active in that region. There is a very real possibility, therefore, that UK companies will be prevented from entering this market in future if commercial conditions improve, because of the recent efforts of their West European and American competitors to 'claim the market', and to occupy a strategically important position for further development if the market begins to improve. The position of UK-owned companies may change, however, as a consequence of strategic alliances with various American, Italian and Japanese power gas turbine manufacturers, and through contacts previously established by subsidiaries in the gas pipeline market. At the time of writing however, Babcock Energy has been recently acquired by Mitsui of Japan; and International Combustion Limited and Parsons Power Generation Systems are currently for sale by Rolls Royce Industrial Power Group.

NOTES

1. *Financial Times East European Energy Report (EEE)*, December 1994, 39/8 ('IVO continues work on St. Petersburg transformation'), provides a project value of approximately US$400 million, but changes were subsequently made in the contract value as a consequence of the increase in value of the Finnish mark against the $US. IVO International's contribution was originally reported as approximately US$180 million, before the increase in value of the Finnish mark against the $US.
2. See 'Combustion Engineering (A)', Harvard Business School Case No.9 - No.9-490-088 (4/5/90) and 'Combustion Engineering (B)', Harvard Business School, Case No.9-490-089 (4/9/90).
3. There is also a Russian report of participation by ABB, Ansal'do and Westinghouse in the two latter projects (*Energetik*, 1994).
4. This information has been extracted from the GEC ALSTHOM 1993-94 Annual Report and converted at the rate of 1 ecu = US$1.15.

REFERENCES

Cooper, J.M. (1988), 'Civilian and military engineering' in Berry, M.J. (ed), *Science and Technology in the USSR*, London: Longman, pp.271-92.

Energetik, (1994), No. 10, pp. 2,3 (A.F.D'yakov).

Financial Times (1995), 1 September, p. 1 ('Mitsui in landmark Europe move').

Financial Times (1997a), 9 April, p. 6 ('Rolls Royce sets up gas turbine joint venture').

Financial Times (1997b), 10 April, p. 10 ('Re-engineering an old name').

Fortune (1994), 'ABB's Big Bet in Eastern Europe', 2 May.

GE Press Release (Feb 1995). 'GE, Kirovsky Zavod Sign Memorandum of Understanding to form Joint Venture for Manufacture, Assembly of Heavy-Duty Gas Turbines for Russian Market.'

GEC ALSTHOM Press Information, 'European Gas Turbine Company signs Co-operation Agreement with Russian Factory', 5 September 1991.

Hill, M.R. and Hay, C.M. (1993), *Trade, Industrial Co-operation and Technology Transfer,* Aldershot: Avebury.

IAEA Bulletin, No. 3/1994.

Imatran voima (IVO) (1994), *Facts in Brief.*

IVO International (1993), *Annual Report.*

Manezhev, S. (1993), *The Russian Far East*, London: Royal Institute of International Affairs, Post-Soviet Business Forum.

Rosengaus, J. (1986), 'The Selection of Steam Parameters for Soviet Thermal Power Plants' in Young, K.E. (ed), *Decision Making in the Soviet Energy Industry: Selected Papers with Analysis*, Falls Church, VA: Delphic Associates, pp. 1-31.

Rosengaus, J. (1987), *Soviet Steam Generator Technology: Fossil Fuel and Nuclear Power Plants*, Falls Church, VA: Delphic Associates.

APPENDIX

Questions for Discussion at Western Companies

1. Is your company currently engaged in commercial arrangements related to electricity generation in the former USSR?

2. (i) If so, could you please give a description of those arrangements, and indicate any further market opportunities which you envisage over the next five years?

    (ii) If not, do you envisage your company becoming engaged in such commercial arrangements in the next five years?

3. Which of your company's present and prospective commercial arrangements have an impact on the reduction of atmospheric pollution in the former USSR?

4. Are you aware of any other power generation or atmospheric pollution problems in the former USSR, which should be addressed in this present research project?

5. Which of your company's present (or future) commercial arrangements in the former USSR require (or will require) technology transfer to that region in design, development, manufacture, installation and operation of electricity generation equipment?

6. What mechanisms and procedures have been (or will be) developed for the transfer of these technologies?

7. Did these mechanisms and procedures for technology transfer include research, design, development, manufacture, installation and operation by industrial organisations in the former USSR?

8. How do you assess the capability of organisations in the former USSR to assimilate fuel treatment, power engineering and power generation technologies, particularly in the area of technologies related to the reduction of atmospheric pollution?

9.   What do you consider to be the barriers to the transfer of power generation technology transfer to the former USSR? Can any of these barriers be removed by multilateral sources available for training and finance?

10.   Have any of your commercial arrangements in the former USSR been related to other commercial arrangements in the former socialist countries of Eastern Europe?

Questions for Discussion at Russian Companies

1.   Where do you expect market opportunities (Russia, other CIS countries, Eastern Europe, Western Europe, North America, Latin America, China, other Asian countries) in the next ten years for the following products:

> gas turbines,
> steam turbines,
> combined cycle systems,
> hydraulic turbines,
> turbine components,
> spares?

(i)   Do you see co-operation with Western companies as an important part of your company's policy to design, develop and produce equipment to meet international environment requirements?

(ii)   With which Western companies do you have co-operation arrangements at present, and in what areas of technology and stage of application (for example, design, production, inspection, implementation)?

(iii) Do you envisage your present range of contacts expanding to other companies?

3.   (i)   Do your company's practices for the documentation of production capability, conformance and traceability match those of Western companies and standards?

(ii)   Are contacts and co-operation arrangements leading to the use of different processes and management procedures than previously?

4.   (i)   Do the practices of your suppliers for the documentation of production capability, conformance and traceability, match those of Western companies and standards?

(ii)   Are contacts with Western companies causing changes in your supplier selection procedures?

5.   (i)   How are the shares of your company divided?

(ii)   How do the shareholders influence the marketing, product, manufacturing, and supply policies of the company?

7. Political, Economic and Commercial Factors

INTRODUCTION

Chapter 6 above has provided information derived from case studies on the activities of a range of power generation, combustion development, power engineering, and energy conservation companies, focusing on the transfer of power engineering technologies and equipment for the reduction of atmospheric pollution in the former USSR. From the case studies of this sample of companies, it is apparent that the volume of of technology transfer has been comparatively low, except for the 'North West' power station project in St. Petersburg involving high levels of participation from Siemens KWU and IVO. Most other companies, however have been concerned with establishing a presence to 'claim the market' in the event of an upturn in sales in the region,

This low volume of technology transfer is also reflected in OECD sales of power generating equipment to the former USSR for the years immediately preceding the commencement of this research project as explained in the next section of this chapter. The volume of business in that region by Western power engineering companies has not matched the apparent need for 'clean' power generation equipment, bearing in mind that many of the technical preconditions are in place to facilitate the assimilation of that technology (see Chapters 4 and 6). The apparent demand has therefore yet to be translated into a viable market, and most Western companies appear to be operating with some caution in this region, whilst still maintaining sufficient presence to 'claim the market' by adopting a defensive strategy against competitors, and await further business opportunities in the region. This chapter will consequently explain the political, economic and commercial factors which influence these strategies, with particular attention being paid to the investment potential for power plant in the former USSR.

WESTERN EXPORTS OF POWER GENERATION MACHINERY

Table 7.1 provides data on the levels of exports of power machinery and equipment from the major OECD countries to the former USSR, and also to the world as a whole from 1987 to 1992. This was the most recent comparative data available at the commencement of the research described in this chapter and Table 7.2 uses those figures to provide information on the share of those exports achieved by each of those same major OECD countries. It is important to bear in mind, however, that these data refer to exports from within the national boundaries of the countries listed, which will include host subsidiaries of foreign multi-national organizations.

Before considering each individual country in turn it is important to note that for the OECD countries as a whole, exports of power generating machinery and equipment to the former USSR has been almost infinitesimal ($322M, or some 0.4 per cent of OECD world exports of these products, in 1992) compared to their exports to other regions, particularly Western Europe ($31bn to the EU in 1992), North America ($20.3bn), and the Far East ($10.7bn including $1.4bn to China) (OECD, 1994). Furthermore, the growth of OECD exports of power generating machinery and equipment to the former Soviet region (38 per cent between 1987 and 1992) has been slower than to other parts of the world (70 per cent between 1987 and 1992 for OECD world exports, and 78 per cent for OECD exports to the EC), but particularly to the Far East including China (105 per cent between 1987 and 1992 for OECD power engineering exports to the Far East, and some 132 per cent to China).[1] These levels of exports are likely to have been influenced by the levels of demand in the market as a consequence of political and economic factors in the former USSR influencing market conditions and investment priorities in that region. These factors are discussed in the following two sections of this chapter.

The case study information presented in Chapter 6 has illustrated the dominant position of German companies and German subsidiaries of multinationals in the supply of power generation equipment to the former USSR, and this dominance is also apparent from published trade statistics for the OECD countries. From Table 7.2, it is apparent that Germany has been the consistent leader amongst the OECD countries in exports of power generating machinery and equipment to the former USSR, varying from 27.8 per cent of the total OECD exports of these products to that region in 1987 to 62.8 per cent in 1991, reflecting that country's traditional position as the leading OECD exporter of engineering products to the former Soviet

Union. The 1991 figure may have been influenced, however, by a recording of contracts completed by organizations in the former GDR, but recorded as 'German' contracts following German unification. Furthermore, the share of German exports of power generating machinery in the former USSR is far higher than its share of OECD countries' world exports of power generating machinery by a factor of more than two, indicating that the former USSR continues as a favoured market for German exporters, who will consequently occupy a strong position if sales opportunities improve in that region. This potential can be met from a position of strength as a major world exporter of power engineering equipment, having a major presence in the large EU market. From the data shown in Table 7.3, it can be seen that the OECD countries' imports of natural gas from the former USSR were some 8-15 per cent of total OECD imports of this commodity between 1988 and 1992, and Germany received between 33 and 49 per cent of those imports from 1988 to 1992. A significant share of German power engineering exports to the CIS export market were partly financed by the sale of Soviet gas in the German market, but particularly the delivery of turbines for pipeline compressor stations, and the countertrade expertise of German companies will continue to assist in the securing of market share in the former Soviet region.

Finland has been the second largest OECD exporter of power generation equipment to the former USSR, with market shares in that region varying from 5.8 per cent in 1991 to 20.2 per cent in 1987. This market share is far higher than Finland's power generation equipment exports to the world as a whole (0.5-0.7 per cent of OECD countries' exports), and the former Soviet region has accounted for more than ten per cent of Finnish exports of these products. OECD Foreign Trade by Commodities statistics also reveal that Finland is a major Western importer of power generating equipment from the former USSR accounting for 35-78 per cent of OECD imports of these products from the CIS countries between 1987-92; and the overwhelmingly major Western importer of CIS electricity accounting for almost 100 per cent of OECD imports of this commodity from the former USSR between 1987-92.

Japan has been the third largest OECD exporter of power generating machinery and equipment to the former USSR from 1987 to 1992, having export sales to the region only marginally lower than those of Finland; although Japan's market share gradually declined from 20 per cent to 5 per cent. Japan's average market shares for those years in the former USSR, however, have approximated quite closely to the 15 per cent share of OECD world exports of power generating machinery and equipment achieved by

Table 7.1 Exports of Power Generating Machinery and Equipment (Selected OECD Countries) (Volume of Exports in $1,000)

	EXPORTS TO FORMER USSR					
Country	1987	1988	1989	1990	1991	1992
Germany	64,644	65,249	101,799	87,165	231,676	119,896
France	9,675	7,043	2,963	4,757	11,131	21,796
Italy	25,257	13,864	8,359	10,082	16,695	10,492
USA	10,073	4,582	12,194	3,015	7,506	38,628
Japan	46,478	39,753	36,607	43,233	30,901	16,240
U.K.	8,797	10,424	4,110	4,013	6,202	33,009
Sweden	957	746	1,589	1,663	1,935	2,696
Switzerland	2,969	2,378	2,572	2,969	3,325	3,441
Finland	47,173	39,846	22,235	44,186	21,437	38,560
OECD Total	232,671	210,730	223,214	256,197	368,634	321,788

Table 7.1 continued

WORLD EXPORTS					
1987	1988	1989	1990	1991	1992
8,460,481	9,119,545	9,591,262	11,195,225	12,214,459	12,753,358
4,179,503	4,887,880	5,043,303	7,801,286	7,825,117	9,001,370
2,314,574	2,190,451	2,512,552	2,844,756	2,954,271	3,106,533
10,359,807	13,049,120	14,165,942	15,569,612	16,967,527	17,949,724
7,454,229	8,721,392	9,512,706	9,888,529	10,745,755	12,443,765
5,326,011	6,920,736	7,747,302	9,398,799	8,964,084	9,717,196
1,231,061	1,503,650	1,484,135	1,710,521	1,602,962	1,737,243
891,388	1,068,842	1,060,307	1,100,464	1,200,729	1,301,175
279,970	307,593	374,232	507,849	512,213	578,568
47,677,799	56,198,565	60,956,409	70,956,680	73,637,478	80,236,918

Source: OECD, *(1994).*

Table 7.2 Market Shares of OECD Countries' Exports of Power Generating Machinery and Equipment

Country	SHARE OF OECD COUNTRIES' EXPORTS TO FORMER USSR (in %)						SHARE OF OECD COUNTRIES' WORLD EXPORTS (in %)					
	1987	1988	1989	1990	1991	1992	1987	1988	1989	1990	1991	1992
Germany	27.8	31.0	45.6	37.9	62.8	37.3	17.7	16.4	15.7	15.8	16.5	15.9
France	4.2	3.3	1.3	1.9	3.0	6.8	8.8	8.7	8.3	11.0	10.6	11.2
Italy	10.9	6.6	3.7	3.9	4.5	3.3	4.9	3.9	4.1	4.0	4.0	3.9
USA	4.3	2.2	5.5	1.2	2.0	12.0	21.7	23.2	23.2	21.9	23.0	22.4
Japan	20.0	18.9	16.4	16.9	8.4	5.0	15.6	15.5	15.6	13.9	14.6	15.5
UK	3.8	4.9	1.8	1.7	1.7	10.3	11.2	12.3	12.7	13.2	12.2	12.1
Sweden	0.4	0.4	0.7	0.6	0.5	0.7	2.6	2.6	2.4	2.4	2.2	2.2
Switzerland	1.3	1.1	1.2	1.2	0.9	1.1	1.9	1.9	2.3	1.6	1.6	1.6
Finland	20.2	18.9	10.0	17.2	5.8	12.0	0.6	0.5	0.6	0.7	0.7	0.7

Source: Calculated from data shown in Table 7.1

Table 7.3 OECD Countries' Imports of Natural Gas ($bn)

		1987	1988	1989	1990	1991	1992
1.	OECD Countries' World Imports	23.023	22.015	20.245	27.341	31.771	29.292
2.	OECD Countries Imports from former USSR	1.835	3.077	2.433	3.708	4.928	4.127
3.	Row (2) ÷ Row (1) (per cent)	8.0	14.0	12.0	13.6	15.5	14.1
4.	German Imports from former USSR	0.008	1.036	1.044	1.615	2.429	1.862
5.	Row (4) ÷ Row (2) (per cent)	0.4	33.7	42.9	43.6	49.3	45.1
6.	Italian Imports from former USSR	0.567	0.780	0.001	-	-	-
7.	Row (6) ÷ Row (2) (per cent)	30.9	25.3	0.04	-	-	-
8.	French Imports from former USSR	0.011	0.041	0.042	0.045	0.038	0.014
9.	Row (8) ÷ Row (2) (per cent)	0.6	1.3	1.7	1.2	0.8	0.3
10.	Austrian Imports from former USSR	0.347	0.316	0.285	0.502	0.508	0.454
11.	Row (10) ÷ Row (2) (per cent)	18.9	10.3	11.8	13.5	10.3	11
12.	Finnish Imports from former USSR	0.154	0.143	0.183	0.237	0.270	0.267
13.	Row (12) ÷ Row (2) (per cent)	8.4	4.6	7.5	6.4	5.5	6.5

Source of import data: OECD, (1994).

that country. It is important to note, though, that Japan was not a significant importer of Soviet energy products over the same time interval, and intergovernmental relations were far from mutually co-operative in view of a continuing dispute between the Japanese and Soviet governments since 1945 over territorial ownership of some of the Kurile Islands. It is probable, however, that the export of Japanese power generation equipment and machinery will increase to the former USSR, if intergovernmental relations improve, followed by government-supported lines of credit. Even if these intergovernmental changes do not occur, however, there is also scope for a significant share of the power equipment and machinery market in the former USSR to be gained by Japanese companies, as a consequence of the import and export trading expertise of the larger Japanese trading houses and the emergence of Japan as the leading exporter of power generation equipment, particularly to the Pacific Rim. At present, though, it appears that Japanese companies still remain averse to risking capital for joint ventures in an environment where the legal and regulatory frameworks remain in a state of rapid change, and exhibit several inconsistencies; but this position may change if Russian natural resources come to be viewed as necessary inputs for further industrial development in the Pacific Rim.

The US share of the power generating machinery and equipment market in the former USSR has traditionally been small compared with its market share in the world as a whole, although there is some evidence to suggest that this situation may be changing as the US share of the power generating machinery and equipment market in the former Soviet region increased from 2.0 per cent in 1991 to 12 per cent in 1992. It is apparent, however, that this market share is far lower than the share of the world market of power generating machinery and equipment held by the US, namely more than 20 per cent. These differences in export performances can be partly explained by the previous political differences between the Soviet and American governments during the years of the Cold War, and associated export control regulations placed on US companies unilaterally by the US government in addition to the multilateral observation of CoCom regulations. As a consequence, US companies often came to be seen as 'unreliable suppliers' by their Soviet customers. In several cases, therefore, US companies have preferred to operate through European subsidiaries or 'manufacturing associates' to serve the Soviet market, which has had a consequential effect on the export performance of the domestic facilities of US companies. This situation appears to be changing, however, as US market shares have been increasing in the former Soviet region, and a major US company (GE) has established a joint venture with the Kirov factory in St. Petersburg.

Although the major concern of this venture will be the provision of turbines for gas pipeline compressor stations, some of the technologies being transferred could also be readily assimilated in power generation.

The share of the power generating machinery and equipment market held by UK exports to the former USSR has also been comparatively low between 1987 and 1991 compared with world market shares of UK exports for these products, although there was evidence of some improvement in 1992 when the UK share of the CIS market increased to 10.3 per cent. A similar pattern is also evident for French exports of power generating machinery and equipment to the former USSR. Information on the market shares of these two countries has been included in view of the existence of a large Anglo-French power engineering international company (GEC ALSTHOM), with large power engineering facilities in the UK and France as well as Germany, and interests in the former Soviet region (see Chapter 6). The differences in international trading context of these two countries is also important, however, as France has been a major importer of natural gas from the former USSR which could be used to parly finance French power engineering exports to that region (see Table 7.3), whereas British imports of this commodity have been negligible as a result of reserves in the North Sea.

The export market share for Italy commenced at 10.9 per cent in the former USSR in 1990, and declined to 3.3 per cent by 1992, which approximates to Italian market shares for power generating machinery and equipment in world markets. This decline in Italian share of the Soviet market, however, is reflected by a decline in Italian purchases of Soviet natural gas, and the sales of the two sets of products are therefore probably closely related. Increases in exports from Italy may increase, however, following reports of large Russian orders recently received by Nuovo Pignone for power engineering products, although these may refer to plant used in natural gas pumping applications rather than power generation.

Market shares are also provided in Table 7.2 for Switzerland and Sweden in view of the existence of a large Swedish-Swiss international company (ABB) which has a significant presence in the former-Soviet market. The export sales volumes for these two countries are comparatively small however, which is not unsurprising bearing in mind the total levels of industrial output from both countries. It is therefore probable that the majority of this company's exports to the former USSR are through its German subsidiaries.

Further data which subsequently became available for 1993, revealed that the country ranking had remained unchanged in terms of volumes of sales of power generation equipment to the former Soviet Union, although Japan had overtaken Germany in terms of the total volume of world exports of power generation equipment ($14.4bn for Japan, compared with $10.8bn for Germany). At the time of writing, OECD data was incomplete for 1994, although available statistics revealed some apparent revisions in reported export volumes of power generation equipment between 1993 and 1994.

POLITICAL INCENTIVES IN THE FORMER USSR

Introduction

The previous section of this chapter has revealed low levels of exports of power generation machinery and equipment during the years immediately prior to the commencement of the research described in this book, and Chapter 6 above has also revealed a comparatively low level of technology transfer related to power engineering. Major factors which have affected these phenonema have been the levels of political and economic incentives to reduce the levels of SOx and NOx in the former USSR, and the priority of these in the allocation of scarce foreign currency for the import of Western equipment for those purposes. This section of the chapter will consequently discuss these Soviet and post-Soviet political and economic factors from domestic and international perspectives, before discussing Western perceptions related to the development of the market.

Soviet and Post-Soviet Domestic Political Priorities

Prior to the establishment of the LRTAP Helsinki Protocol for SOx reduction in 1985, and the Sofia Protocol in 1988 (see 'External Political Pressures...' below), there was little evidence of any internal political pressures within the former USSR to reduce the levels of atmospheric pollution within that region. Decisions on environmental pollution rested mainly with industrial ministries, who utilised the majority of their resources to increase output rather than to meet environmental objectives (Whitefield, 1993, pp. 158-62) as a consequence of the output-related incentives of the centrally planned economy. Furthermore, the Soviet censorship system limited the open discussion of environmental concerns, and hindered the development of an environmental political movement.

The *'glasnost'* policies of the Soviet government introduced in 1985 permitted a more open discussion of the environmental problems faced by the former USSR and facilitated the growth of an environmental political lobby. This process was reinforced by the information gathered on the scale of radiation following the Chernobyl accident and from previous nuclear testing locations in Kazakhstan. In addition, a State Committee for Environmental Protection *(Goskompriroda)* was established in 1988, to submit proposals to the State Planning Committee *(Gosplan)* for environmental protection measures to be included as targets in the five year plans for industrial ministries and their subordinate factories. Each republic was then to establish counterpart committees with subordinate offices in municipalities and local regions. Following the reform of the parliamentary system in 1989, a Committee on Ecology and the Rational Use of Natural Resources was created with responsibility to the USSR Supreme Soviet (IMF et al., 1991, p. 12).

The implementation of measures proposed by *Goskompriroda* was hindered by bureaucratic inertia, difficulties in the recruitment of adequate staff, and continuous tensions with industrial ministries over powers and responsibilities, together with concerns that pollution control could lead to higher costs and consequent plant closure and unemployment. *Goskompriroda* was also advocating the introduction of economic incentives based on fines and the use of the pricing mechanism, but the former was opposed by industrial ministries, whilst the latter was difficult to introduce because of the lack of a realistic pricing system from which to start (IMF et al., 1991, p. 12). Furthermore, although some 2 per cent of total state investment was allocated to environmental expenditures in 1990, which represented a 25 per cent increase from the 1988 budget, it was questionable whether the level of investment was sufficient for the scale of the problem. A major backlog existed from the lack of environmental concerns in the previous central planning system, and also as a failure to meet 1988 targets since only 55-65 per cent of air and water pollution targets were achieved (IMF et al., 1991, p. 14).

One positive feature of environmental protection in the former USSR, however, was the approval of a state standard on NOx emission limits for boilers (GOST 28269-89) drafted partly as a consequence of the Soviet signing of the 'Sofia Protocol' in 1988. Another set of data were drafted at about the same time which set limits for boiler emissions from equipment supplied with gas cleaning equipment by the Ministry of Heavy Engineering of the USSR *(Mintyazhmash SSSR)* to the Ministry of Power of the USSR *(Minenergo SSSR)*. These limits were subsequently approved by *Goskompriroda* in 1989. A comparison of the information provided for NOx emissions in both sets of data showed that variations existed between the state standard and the *Goskompriroda* approved emission limits (see Chapter 2), although that was not surprising as GOST 28269-89 referred to the acceptance testing requirements for new boilers including those to be installed in power stations, rather than power station emissions in practice. This comparison of the data also showed the necessity of fitting post-combustion equipment for large boilers burning heavy fuel oil and solid fuels, particularly those burning hard coal. The permissible *Goskompriroda* NOx limits may also have been achievable by the use of low NOx burners, and thereby achieved in a more cost effective way at the combustion stage, but the widespread implementation of low NOx burners would have required reductions in the allowable NOx emission limits from the boilers themselves as defined by GOST 28269-89.

There is little evidence to suggest that atmospheric emissions from combustion processes have been reduced in the former constituent republics since the fragmentation of the USSR in 1991, although the Russian Federation has recently introduced new legislation for SOx and NOx emissions from new, but as yet unbuilt, coal-fired power stations (see Chapter 2). Reductions in aggregate levels of SOx and NOx emissions have therefore occurred as a consequence of the fall in industrial output and associated demand for electricity, and fuel switching from coal and oil to natural gas. In addition, each republic has had to establish its own system of government which has had consequences for the opportunity and time to draft environmental legislation, although the Russian Federation was able to assimilate most of the former Soviet governmental structure because of its location in Moscow (including the establishment of a Ministry of the Environment, *Minpriroda*, in place of *Goskompriroda*) and hence had the opportunity to draft legislation at a faster pace than elsewhere in the former Soviet Union. These advantageous circumstances were not fully realised, however, as Russia faced constitutional problems related to the relative powers of President and Parliament which hindered the introduction and implementation of a wide range of legislation including the enforcement of environmental limits. Finally, the severe economic constraints created by budget and foreign trade deficits (a $42bn foreign trade debt was inherited by the Russian government from the former Soviet Union) together with extremely high rates of inflation (Rutland, 1992; Sakwa, 1993), have acted as severe limitations to investment in environmental control equipment, or the import of such equipment from the West.

Decision-makers in the former USSR currently face stark choices of either doing nothing to reduce atmospheric pollution, or using scientific resources and scarce foreign currency which could be used for other applications. It is almost impossible to estimate the investment required for the reduction of atmospheric pollution in the former Soviet Union, although estimates varying from some $6-8bn to some $20bn over the next 15 years have been provided for Poland which has an installed coal-fired capacity of 29,000MW.[2] It is likely that the installed coal-fired capacity in the former USSR is almost double the Polish figure, and some $12-40bn may consequently be required for pollution control of coal-fired power stations in that region over the next 15 years. In view of the importance of pollution to the international community, however, and the other demands being placed on reserves of hard currency in the former USSR, it is difficult to see how financial resources for the reduction of emissions can be obtained, without access to unilateral and multilateral sources of finance.

Power Station Specific Emissions

From the data shown in Table 7.4 based on data previously presented in Chapter 2 above, it appears that power station specific emissions of SOx and NOx (that is, emissions per unit of electricity generated) were lower for the former USSR than for the UK and Germany, in the selected base years of 1980 and 1987 respectively, except when the upper estimate for SOx emissions is taken for the former Soviet region and compared with the German figure. It is important to bear in mind, however, that the data shown in Table 7.4 are far from precise in view of the assumptions made to estimate the SOx and NOx emissions, and the differences in proportions of fossil-fuels and nuclear energy used to generate electricity within the three countries chosen for comparison. It is also important to keep in mind that all three countries have attempted to reduce their levels of atmospheric emissions since 1980, with both British and German power stations making substantial investments to reduce the levels of pollutant atmospheric emissions.

From these data though, it is apparent that there is little evidence to suggest that the average levels of power station specific emissions of SOx and NOx per KWh of electricity produced, have been higher for the former USSR than for other industrialised countries, partly because of the comparatively high levels of use of natural gas and hydro-power (see Chapter 5). There are therefore no pressing reasons for the governments of the former USSR to create political incentives for an aggregated approach to the reduction of specific atmospheric emissions from power stations. Attention may be paid to energy efficiency with a consequent (although not direct) reduction in fuel consumption and associated atmospheric emissions,

Table 7.4 Specific Emissions of Atmospheric Pollutants from Power Stations (*Early 1980s*)

	Production of Electricity (TWh) (1980)	SOx Emissions (M tonnes) (1980)	Production of Electricity (TWh) (1987)	NOx Emissions (M tonnes) (1987)	Specific Emissions (g/kWh of SOx) (1980)	Specific Emissions (g/kWh of NOx) (1987)
UK	250[1]	3.8[3]	260[1]	0.8[3]	15.2[5]	3.1[6]
Germany	300[1]	2.2[3]	350[1]	0.8[3]	7.3[5]	2.3[6]
Former USSR	1294[2]	7.0-10.6[3]	1665[4]	2.7[3]	5.4-8.2[5]	1.6[6]

Notes:
1. See Boehmer-Christiansen and Skea (1991), pp.140-1.
2. See footnote to Table 2.2.
3. See Table 2.13 above.
4. Goskomstat (1988), p. 120.
5. Calculated from data in columns 1 and 2 (for example, 15.2 g/kWh = 3.8M tonnes/250 TWh).
6. Calculated from data in columns 2 and 4 (for example, 2.3 g/kWh = 0.8M tonnes/260 TWh).

but such a strategy is more likely to be driven by increases in energy prices than political objectives. Alternatively, political attention is more likely to be directed towards local problems with some international consequence. An example of such a project is the 'North-West' power station in St. Petersburg referred to in Chapter 6. Investment in this gas-fired power station could be viewed as a means to achieve reductions in atmospheric emissions in North West Russia, to meet the requirements of an agreement with the Finnish government to reduce SOx emissions in the Leningrad, Murmansk and Karelian regions. A reduction of 50 per cent in 1995 compared with 1980 was specified in this agreement, however (Lyalik and Reznikovskii, 1995, p. 64), and so this project can only be assessed in the context of future programmes of SOx reduction, combined with the possibility of replacing nuclear power generation in North West Russia, which is a topic of major environmental concern to the Finnish population.

Local Levels of SOx and NOx Emissions

The estimates reported in Chapter 2 above have provided information on the aggregate levels of SOx and NOx emissions for European Russia, the Russian Federation, European regions of the former USSR and the USSR as a whole. These aggregate levels have been seen to be high when compared with other countries on the European continent. These high levels are to be expected, however, in an industrialized country with a large population and harsh winter climate creating energy demands for industrial and domestic purposes, and a large land mass with associated long distances for distribution. If the levels of SOx and NOx emissions are considered in aggregate, and divided by the populations and land areas of UK and Germany as two typical European countries (see Table 7.5), then it becomes apparent that the levels of density of SOx emissions for the former USSR were double those of Germany and higher than those of UK for SOx emissions per capita, but marginally lower than those two countries for NOx emissions per capita as a result of the comparatively undeveloped Soviet load transport infrastructure. These values of per capita SOx and NOx emissions are higher than those presented in Table 2.10 above, however, as that table refers to stationary sources only using government-reported emission levels, whereas Table 7.5 includes a compensation for apparent under-reporting of SOx levels (see Table 2.5) and non-stationary sources of NOx emissions (see Table 2.9). As a result of the large land area occupied by the former Soviet Union, however, the SOx and NOx emissions per unit of land area are far lower than for Germany and UK, even when using the higher estimated values of Soviet SOx and NOx emissions.

Appendix D of the EMEP report (Tuovinen et al., 1994, pp. 24-6) noted high levels of oxidised sulphur deposition per unit of area in several European regions of the former USSR during the 1985-1993 time interval (for example, eastern Ukraine $3.6g(S)/m^2$, North West Russia [Kola and Karelia] $1.7g (S)/m^2$, Estonia $1.1g(S)/m^2$); although these figures were lower than for some other Western and Eastern European countries (for example, $5.1g(S)/m^2$ in certain regions of the UK, $8.6g(S)/m^2$ for some regions of the former German Democratic Republic, and $5.2g(S)/m^2$ for some regions of Poland).

For depositions of oxidised nitrogen, the levels in central Russia and Eastern Ukraine were in the region of $0.24g(N)/m^2$, compared with some $0.5g(N)/m^2$ for some certain regions of Germany and the UK. From these data, therefore, there is an apparent need for SOx emissions to be reduced to benefit the health and well-being of the population, particularly in those cities where there are particularly high levels of emissions per capita as listed in Chapter 2. The low levels of aggregate SOx and NOx emissions per unit area of territory are therefore unlikely to be a cause of great concern. There are specific regions, though, (for example, Kola and Karelia, and Estonia) where the levels of sulphur deposition are far higher than those of their Nordic neighbours for sulphur deposition (for example some five times higher than that of Finland) and therefore remain a cause for concern in the Scandinavian region as explained in the following section of this chapter.

External Political Pressures from Multilateral Agencies

United Nations Economic Commission for Europe (ECE) and Long Range Transboundary Air Pollution (LRTAP) Convention
(Boehmer-Christiansen and Skea, 1991, pp. 27-30)
The ECE has been a long-standing proponent of reduced atmospheric pollution, and is an institution which has contained the USA, Canada, the former USSR, and Western European countries within its membership. The questions of energy and pollution were placed on the agenda for this international body following a speech by President Brezhnev in 1975 related to the Helsinki Agreement on Security and Co-operation in Europe. In this speech, President Brezhnev claimed that both East and West shared common problems in environment, energy and transport.

A 'Long Range Transboundary Air Pollution' (LRTAP) Convention was consequently established within the institutional framework of ECE, and this Convention subsequently established the Helsinki Protocol in 1985. Most of the countries within ECE committed themselves to observe this protocol by reducing either SOx emissions or transboundary flows of SOx pollution by 30 per cent of the 1980 level, by the year 1993. A second SOx protocol was signed in Oslo in June 1994, relating to targets (or 'national budgets') for the years 1995, 2000, 2005 and 2010. In addition, a Sofia Protocol was signed in 1988, which required a freeze in NOx emissions from 1994 onwards, at the 1987 level. Negotiations are presently under way for a new Protocol which is intended to cover multi-pollutants and multi-effects.

From the data presented in Chapter 2 above, it appears that the European region of the Russian Federation was apparently on target to meet the 1985 Helsinki Protocol requirements for SOx reduction assuming that there was no major increase in SOx emissions in 1993 compared with the 4.4M tonnes figure provided for 1992 (see Table 2.3). Later data on emissions from the European region of Russia provided a figure of 3839 thousand tonnes of SO_2 for 1992 (ECE, 1994, Table 1), further reinforcing the view that the region achieved a reduction of far more than 30 per cent (that is, some 46 per cent) compared with 1980 (7171 thousand tonnes). Significant reductions far in excess of 30 per cent (that is, 48 per cent) compared with 1980 were also reported for Ukraine, which reduced its 1980 SO_2 emissions figure of 3850 thousand tonnes to 2376 thousand tonnes in 1992 (see Table 2.3), and 2194 thousand tonnes in 1993 (ECE, 1994, Table 1).

Furthermore, it appears that the European region of the Russian Federation was also just on target to meet the requirements of the 1988 Sofia Protocol during 1993 using data provided for 1987 and 1992 NOx emissions (2353 thousand tonnes and 2326 thousand tonnes respectively), provided that NOx emissions did not increase in 1993. Revised data from ECE (1994) suggest that the Russian Federation is even more likely to meet the Sofia Protocol than the data in Table 2.7 suggest, as NOx emission levels of 2653 thousand tonnes and 2298 thousand tonnes are provided in that document for European Russia in 1987 and 1992 respectively (ECE, 1994, Table 2). In the case of Ukraine, however, NOx reductions in 1992 were some 24 per cent lower than in 1987, according to the data presented in Table 2.6.

From these data, therefore, there do not appear to be any grounds for further major ECE multilateral pressures on the Russian and Ukrainian governments to to define the reduction of SOx and NOx emissions at the aggregate level as major domestic priorities, and to consequently increase the assimilation of Western technologies in the field of reduction atmospheric pollution, unless NOx and SOx emissions increase as a consequence of expansion in industrial activity. Pressures for SOx and NOx reduction are therefore more likely to come from other political and economic sources, as outlined below.

The 1991 European Energy Charter (Fraser, 1992, pp. 40-41)
The 1991 European Energy Charter is a political commitment between Europe, the member states of the CIS, USA, Japan, Canada and Australia, to establish greater co-operation in the international energy market in Europe, and the establishment of basic international regulations for the energy industry. Environmental protection is included as part of one of three main areas of future action, and is also one of the twelve topics to be negotiated within a legally binding Basic Agreement. It is possible, therefore, that the countries comprising the former USSR will be subject to external European pressure through this Charter to reduce the levels of environmental pollution in conformance with European and other international norms. This could occur in the European market for electrical power, or for those products consuming electricity as a major element of production costs, such as non-ferrous metals (see 'External Pressures from West European Governments' below).

The European Commission Large Combustion Plant (LCP) Directive, July 1988
In the 1970s and early 1980s, each European country followed its own code of practice with regard to acceptable levels of emission, but during the late 1980s the European Commission attempted to standardise the emissions of SOx and NOx from new large power stations. The units chosen were those of milligrams (mg) of pollutant per normalised cubic metre (Nm^3) of emission, according to defined conditions of pressure, temperature and oxygen content. For plants of thermal capacity greater than 500MW burning solid fuel, the NOx limit was set in the June 1988 LCP Directive at 650 mg/Nm^3, or 1300mg/Nm^3 for solids with less than 10 per cent volatile compounds; 450mg/Nm^3 for liquid fuels and 350mg/Nm^3 for gaseous fuels. For SO_2, the limit was set at 400 mg/Nm^3 (IEA, 1988, pp. 163-166); Boehmer-Christiansen and Skea, 1991, pp.236-7). For existing large power stations, each country agreed to reduce emissions (or 'budgets') for both sulphur dioxide and nitrogen oxide. In the case of the UK for example, within a 1980 SOx total emission figure of 4.67M tonnes, 3.88M tonnes of emissions from large combustion plants were targeted to be reduced by 20 per cent by 1993, 40 per cent by 1998 and 60 per cent by 2003. For NOx emissions a 1987 adjusted emission figure of 1.02M tonnes from large

combustion plants was expected to be reduced by 15 per cent in 1993, and 30 per cent in 1998. For the Federal Republic of Germany, LCP SOx levels of 2.23M tonnes were to be reduced by 40 per cent in 1993, 60 per cent in 1998, and 70 per cent in 2003. NOx emissions were expected to be reduced from a 1987 adjusted level of 0.87M tonnes by 20 per cent in 1993 and 40 per cent in 1998 (Boehmer-Christiansen and Skea, 1991, pp. 238-41).

It is important to note, however, from the discussion in the previous subsection on the LRTAP Convention, that Russia's and Ukraine's SOx emissions in 1992-3 had reduced by more than 40 per cent since 1980, and Ukraine's NOx emissions in 1992 were 24 per cent less than in 1987. These reductions therefore appear to meet many of the targets laid down for West European countries through the LCP Directive; although they have been mainly achieved as a consequence of decline in industrial output in the former USSR rather than the implementation of cleaner combustion processes except in the cases of fuel switching to gas. In addition, new power stations constructed to the Russian normative documents and standards outlined in Chapter 2 would appear to meet the LCP Directive's SOx and NOx emissions requirements for new plant. Evidence in Chapters 3 and 4, however, on fuels and combustion processes used in the former USSR illustrates that further technical development is required to consistently meet all of the requirements of contemporary standards and normative documents.

At this point in time, it is difficult to predict the extent to which the requirements of the LCP Directive will be applied to the former USSR, as the constituent states of the former USSR are not members of the European Union, and more attention is likely to be paid to Poland, Hungary and the Czech Republic in view of their more immediate objective to become members of the European Union. Western Europe may be viewed as a market for electricity produced in the former USSR, however, to extend the volumes of this commodity already sold into Finland. Pressure may be consequently placed on power stations in the former USSR located near to the European border to conform to European standards to preclude claims of unfair competition over electricity exports to the European Union, and the construction of the 'North-West' (see Chapter 6) combined cycle power generation plant may be seen in that context. Similar pressures may also become apparent for exportable products manufactured by energy-intensive processes and for which electricity is also a significant component of total production cost, particularly as some multilateral financing organizations (such as the European Bank of Reconstruction and Development) require an environmental appraisal to be carried out as a pre-requisite to investment approval (EBRD, undated, pp. 10, 11). The multilateral political pressures capable of being exerted by the European Union can therefore be viewed as being potentially far stronger than those of the ECE, in view of the economic incentives that can be exerted through EC regulations, and the possibilities for access to the large West European market.

External Political Pressures from West European Governments

Transboundary transfers of SOx and NOx emissions

In spite of the large volume of pollutants emitted by Soviet industry, the former USSR has been a net importer of oxides of sulphur and nitrogen from Europe as a whole. This is particularly the case for emissions from Poland, East Germany, Czechoslovakia, Hungary, and West Germany, (see Table 7.6), and the governments of the various countries comprising the former USSR may therefore hold the view that some of their neighbours have a worse record than themselves in terms of transboundary pollution.

Particular concern over the levels of atmospheric pollution transferred to the former USSR exist in some of the Scandinavian countries, however, because of the overall effects of imported pollution on those countries' natural resources. This is particularly the case for Finland, where the European region of the former Soviet Union was the source of some 30 per cent of oxidised sulphur and almost 15 per cent of oxidised nitrogen deposited on that country between 1985 and 1993 (see Table 7.7), and the corresponding figures for oxidised sulphur depositions for Sweden and Norway are 8.7 per cent and 9.1 per cent, respectively.

When the transfers of emissions from the Scandinavian countries to the former USSR are included to arrive at a level of net transboundary transfers, Finland's, Sweden's and Norway's net proportion of Soviet emissions in total depositions of SOx remain high at 15.5 per cent, 5.0 per cent and 8.0 per cent respectively; although Norway emerges as a net exporter of SOx to the former USSR, and all of the Scandinavian countries emerge as net exporters of NOx to the Soviet region. If the transboundary movements of emissions from the former USSR are underestimates, however, in similar proportions to the official data for Soviet emissions cited in Chapter 2 above, it is possible that the transboundary movement of SOx and NOx from the former USSR to the Scandinavian countries could be considerably higher

Table 7.5 SOx and NOx Emissions per Capita of Population and per Unit of Land Area (UK, Germany and USSR)

	Area (thous km^2) (Col 1)	Population (1980) (millions) (Col 2)	SOx emissions (1980) (M tonnes) (Col 3)	Population (1987) million (Col 4)	NOx emissions (1987) (M tonnes) (Col 5)	SOx emissions g/m^2 (1980) (Col 6)[8]	SOx emissions tonnes per thousand capita (1980)[8] (Col 7)[9]	NOx emissions g/m^2 (1987) (Col 8)[10]	NOx emissions tonnes per thousand capita (1987) (Col 9)[11]
UK	242[1]	56[1]	5.3[2]	56[1]	2.6[3]	20.2	94.6	10.7	46.4
Germany	249[4]	62[4]	3.6[2]	61[4]	3.1[3]	12.9	51.6	12.4	50.7
Former USSR	22,402[5]	264[5]	28.5[6]	283[7]	11.2[6]	1.3	108.0	0.5	41.9

Sources:

1. UK Office of Population and Censuses
2. See Table 2.1.
3. See Table 2.6 (footnote).
4. Statistical Yearbook, Federal Republic of Germany, 1994.
5. Estimated from 1979 figure of 262M provided in Goskomstat (1990).
6. See Tables 2.5 and 2.9.
7. Estimated from 1990 figure of 289M and 1989 figure of 287M (see Goskomstat [1990], p.17).
8. Col 6 = Col 3 / Col 1.
9. Col 7 = Col 3/Col 2 × 1000.
10. Col 8 = Col 5 / Col 1.
11. Col 9 = Col 5/Col 4 × 1000.

Table 7.6   Average Annual Net Depositions of Oxidised Sulphur and Nitrogen in the Former USSR (1985-1993: 100 of Tonnes per Annum)

	Oxidised Sulphur Depositions			Oxidised Nitrogen Depositions		
Poland - USSR	3154 - 18	=	3136	1022 - 70	=	948
East Germany - USSR	1972 - 18	=	1954	263 - 7	=	256
Czechoslavakia - USSR	1133 - 41	=	1092	443 - 15	=	428
Romania - USSR	900 - 327	=	573	302 - 92	=	210
Hungary - USSR	623 - 33	=	590	113 - 12	=	101
West Germany - USSR	351 - 22	=	329	748 - 9	=	739
Europe - USSR	17,873 - 4,759	=	13,114	7,241 - 1,159	=	6082

Note: The left hand figure in each column records transfer of pollutant into the USSR, and the middle figure in each column records the transfer from the USSR. The right hand figure in each column consequently records the net transfer of each pollutant into the former USSR for each of the listed countries.

Source: Tuovinen, J.-P. and Barrett, K. Styve, H. (1994), *Transboundary Acidifying Pollution in Europe: Calculated fields and budgets 1985-93.* Det Norske Meteorologiske Institut Technical Report No.129 (EMEP/MSC-W Report 1/94), July 1994, Appendix E.

than those provided in Table 7.7, and that Finland and possibly Norway may emerge as net importers of NOx from the former Soviet region.

It is important to bear in mind, however, that particular power generation plant and industrial facilities emit high concentrations of atmospheric pollutants which are then transferred to Finland at high levels of density, and these local transfers can have a more significant effect in a region than is suggested by national transboundary data. Particular examples of sources of high sulphur transboundary pollution cited by the Finnish authorities are the Pechenganickel and Severnonickel combines in the Murmansk region, which have deposited some 10,000 to 15,000 tonnes of sulphur annually in Lapland as a consequence of the use of high sulphur ores (20-30 per cent sulphur content) from Norilsk combined with the building of high stacks; the Kostamuksha metals and iron pellet plant, Segezha and Kondopoga pulp and paper mills, and the Petrozavodsk power plant in Karelia; the Kirishi oil-fired power plants and Svetogorsk paper mill in the Leningrad region; and the Eesti and Baltai power plants in Narva, Estonia which use low calorific oil shale including up to 50 per cent sulphur and inflammable minerals, and account for some 10-20 per cent of the total sulphur depositions in Southern Finland (Hiltunen, 1994, pp. 13-15).

It is more likely, therefore, that pressure to reduce levels of atmospheric pollution in the former USSR will be exerted by neighbouring countries such as Finland, and there is evidence that this pressure is also being supported by technical and economic assistance. Some F40m marks (approximately £7m) have already been allocated by the Finnish government as aid for the reduction of pollution in North West Russian, the Kola Peninsula and the Baltic States, provided that 50 per cent of the pollution control investment is made by an organization in the former USSR (Hiltunen, 1994, pp. 13-15). In the previous section of this chapter attention was drawn to the role of the EU as a centre of influence on political decision-making for the reduction of atmospheric pollution in the former USSR, particularly where environmental costs may influence prices. It is probable, for example, that the existence of polluting smelters and foundries featured in discussions between the Russian Federation and the EU, in relation to quotas for Russian aluminium exports to Western Europe (as explained in the following sub-section); and similar concerns may be extended further down the supply chain to include energy inputs, particularly where energy is an important cost component.

If multilateral funds are obtained to reduce levels of atmospheric emissions they may be more effectively spent in the purchase of licences and

Table 7.7 Sources of Deposits of Oxidised Sulphur and Nitrogen in Scandinavia (Annual Averages 1985-1993)

Country in which emissions are deposited (Col.1)	Depositions of Oxidised Sulphur (tonnes of S per annum x 100)					Depositions of Oxidised Nitrogen (tonnes of N per annum x 100)				
	from former USSR (Col.2)	to former USSR (Col.3)	deposition including domestic deposition (Col.4)	Col(2)/Col(4) % (Col.5)	[Col(2)-Col(3)]/Col(4) % (Col.6)	from former USSR (Col.7)	to former USSR (Col.8)	total deposition including domestic deposition (Col.9)	Col(7)/Col(9) % (Col.10)	[Col(7)-Col(8)]/Col(9) % (Col. 11)
Finland	553	267	1846	30%	15.5%	108	223	761	14.2%	-15.1%
Sweden	199	85	2299	8.7%	5.0%	69	206	1291	5.3%	-10.6%
Norway	129	15	1417	9.1%	8.0%	19	53	864	2.2%	-3.9%
Denmark	8	105	595	1.3%	-16.3%	3	144	270	1.1%	-52.2%

Source: Tuovinen, J.-P. Barrett, K. and Styve, H. (1994), *Transboundary Acidifying Pollution in Europe: Calculated fields and budgets 1985-93.* Det Norske Meteorologiske Institut Technical Report No.129 (EMEP/MSC-W Report 1/94), July 1994, Appendix E.

technology rather than in the purchase of assembled such equipment, as the readily available, but presently under-utilised skills and manpower which exist in the Russian Federation and Ukraine could thereby be utilised in the manufacture of such equipment. In the 1993 EU TACIS Programme, for example, a budget of some 1.7 MECU was earmarked to support the expansion of modern combined cycle generating capacity in St. Petersburg. These funds were designated for analysis, advice, and training; and the purchase of equipment for training and management systems (TACIS Contract Information Budget 1993). In 1994, 1.5 MECU was earmarked for the stabilisation of operating regimes at Russian power stations, and 1.0 MECU has also been earmarked for the development of boilermaking production capacity in Ukraine, with an additional 2 MECU for reform of the energy sector and 2.7 MECU for studies of power generation efficiency in that country (TACIS Contract Information Budget 1994).

European environmental pressures may also be exerted on constituent republics of the former USSR through the framework of the European Energy Charter, particularly if these republics seek to increase their share of the European market for electricity, or to expand their sales of those products consuming electricity as a major element of production costs, such as metal products. It is quite probable, however, that Western European pressures for pollutant control will be higher for the Baltic States, and Poland, Hungary, and the Czech and Slovak republics in view of their location, and their wish to become full members of the European Union. Such reductions would also have a beneficial effect for the former USSR, however, as that region has been a net importer of SOx and NOx from Eastern Europe (Tuovinen et al., 1994, Appendix E).

ECONOMIC INCENTIVES IN THE FORMER USSR

Environmental Factors and Russian Exports

There are clear international economic advantages to be gained by the former USSR from the reduction of atmospheric emissions of SOx and NOx into the atmosphere, with associated production of acid rain and deleterious effects on woodlands and forests. The former USSR has been a major exporter of timber to the OECD countries with exports increasing from $1.4bn in 1987 to $1.9bn in 1989, before decreasing to $1.3bn in 1992. These levels of exports to the OECD countries were only surpassed by USA, Canada, Finland, Sweden and Malaysia, and the USSR supplied some 4.6

per cent of OECD imports of timber in 1992, and some 6.9 per cent of imports in 1991. The continuity of deliveries of timber to the West is contingent upon a range of agricultural, production and distribution factors, but the future availability of good quality forests free from the damage of acid rain is a major factor. Furthermore, the former USSR has been a significant exporter of non-ferrous metals to the OECD countries, exporting some $3.5bn in 1991 (supplying some 6.25 per cent of total OECD countries imports of this commodity) and some $2.9bn in 1992. Exports of iron and steel also accounted for some $1bn in 1992, although the OECD market share (at some 1.6 per cent) was far lower than that for non-ferrous metals. In recent conditions of aluminium supply exceeding short term demand, however, and the consequent effect on price, concerns were expressed by Western manufacturers and governments about the high levels of production in the former USSR and exports to the West, at a time when Western manufacturers were restricting and reducing output (*Financial Times*, 13 January 1994, p. 5). Limits were subsequently agreed on aluminium exports from the former USSR to the European community (*Financial Times*, 28 October 1993, p. 38) and the Technical Assistance to the CIS (TACIS) Programme made funds available for various environmental and energy conservation projects in the non-ferrous metals industry in that region (TACIS Contract Information Budget, 1993; TACIS Contract Information Budget, 1994).

Although the present author has been unable to discover any official statement which explicitly linked these two policies, it is not improbable that the effects of lower environmental control on reduced production costs in the former USSR were of great concern to Western manufacturers during the 1993/94 decline in prices. The emission of hazardous substances from CIS facilities, for example, was estimated to be some eight to ten times higher than that allowed by European regulations, with associated effects on production costs (*Financial Times*, 13 January 1994). The proposed European arrangements of restricted exports and technical assistance in environmental control is therefore one approach to the resolution of these problems.

In the case of American industry, however, one US industrial spokesman recommended the storage of excess aluminium production from the former USSR as future collateral to the release of Western funds to assist in the restructuring of the aluminium industry in that region, and investment in pollution control equipment (*Financial Times*, 6 January 1994). Although Western concerns have been directed at smelters and foundries in the first instance, subsequent concerns could also be directed at 'unfair competition' through cheap energy (even though electricity prices are becoming more liberalised in the former USSR) which in turn could be influenced by a lack of concern over energy-related environmental factors in the former Soviet region.

Energy Conservation and Russian Exports

Reduction of atmospheric pollutants through energy efficiency is another course of action available to Russian policy-makers, particularly through consequent reductions in fuel consumption per unit of power generated. The previous Soviet policy of centrally-driven fuel efficiency through reduced fuel allocations per unit of power generated, achieved fuel savings per unit of electrical output from almost 600 grams of 'conventional' fuel (rated at 7,000 kcal/kg) per kilowatt hour of electricity in 1950, to less than 350 grams of conventional fuel per kilowatt hour in 1980, with economies of scale accounting for a significant proportion of these savings (Rosengaus, 1986, p. 12; Hewett, 1984, p. 116). Such savings in fuel consumption should have had a significant effect of atmospheric pollution, but this relationship is difficult to gauge without detailed information of the mix of fuels and combustion conditions used.

Pressures for reduced fuel consumption are also likely to continue during the transition to a market economy, particularly as privatization and price competition are introduced into the power generation sector and state subsidies are progressively removed from fuel prices. Increases in operational efficiency are particularly important in the present environment of 'energy crisis' caused by reduced levels of production of oil and coal in the early 1990s compared with the late 1980s, as a result of difficulties in extraction and transport and lower levels of demand, with consequent increases in costs per unit of output as economies of scale have been reduced. Natural gas production has continued to increase, however, albeit in smaller increments since 1990 because of problems in extraction and transmission, and reduced levels of demand particularly from Eastern Europe. It is doubtful, however, whether continued fuel savings will be available in power generation from economies of scale, since reductions in demand for electricity may cause large plants to operate at sub-optimal levels of capacity. Furthermore, the scope for the rationalisation of power station capacity may be limited by the necessity of retaining particular co-generation facilities to provide district heating. Fuel savings will therefore be sought from improved combustion processes and efficiencies in power station operating procedures and these demands for savings will be increased if fuel prices continue to rise, particularly if fuel prices rise in real terms.

Increased domestic efficiency in the former USSR is also likely to be beneficial to export options, as the region (particularly Russia) is heavily dependent upon the sale of energy products for export earnings. In 1990 and 1991, for example, the CIS sold some $3.7bn. and some $4.9bn of gas respectively accounting for some 13.5 per cent and 15.5 per cent of those countries' gas imports during those years: and in 1990 the former USSR sold some $17.3bn. of mineral fuels, lubricants and related materials to the OECD countries (6.3 per cent of OECD countries' imports) and some $16.2bn. in 1991. Subsequent data for 1992 and 1993 shows a decrease in the levels of OECD imports from the former USSR for both of these commodities ($4.1bn. of natural gas in 1992 and $2.2bn in 1993; $14.9 of mineral fuels, lubricants and related materials, and $12.5bn in 1993), although some of these reductions may have been due to OECD's re-evaluation of post-1991 exports from the former Soviet region. If there is consistency in the evaluation of these exports, however, the CIS countries as a group lost market share in OECD imports of both gas and mineral fuels between 1990 and 1993. Although the 1990-1993 consistency in data should be treated with some caution, the smaller volumes of exports to the OECD countries of CIS energy commodities may be increased by attention to energy conservation programmes in order to reduce the effects of potential bottlenecks in supply.

Investment in Power Plant

One of the major factors affecting future business opportunities for power engineering equipment in the former USSR is the level of demand for power generation capacity in that region. From material presented earlier in this chapter, it has been concluded that the multilateral political and economic pressures likely to be placed on the former USSR to introduce 'cleaner' combustion equipment to reduce SOx and NOx emissions will be comparatively low: Russia and Ukraine appear to have met the ECE's requirements for reduced SOx and NOx emissions; partly as a consequence of reductions in demand for power because of industrial recession in that region but partly through continued switching from coal and oil to natural gas. Furthermore, the majority of Russian coal-fired power generation capacity is located in the Urals and Siberia, which are long distances from Europe with its associated concerns over sulphur depositions and political pressures for their reduction, although pressures may be exerted to reduce CO_2 emissions from coal firing (see Chapter 8).

The incentive for investment in cleaner burning power generation equipment in the former USSR is therefore more likely to be economic rather than political. As explained in Chapter 5, a significant proportion of generating capacity is old and obsolete, and will require replacement even if industrial decline continues in the region. Natural gas can burn with a higher level of efficiency (some 38 per cent efficiency from gas-fired plants) than coal (less than 30 percent in many cases); and combined cycle systems can burn at even higher efficiencies enabling some 15-20 per cent in savings to be made in fuel savings, and up to some 40 per cent reductions in space and pay back periods (50-55 per cent) (see Chapter 4). The economic arguments for natural gas, particularly in a combined cycle configuration also lead to the potential for reduced SOx and NOx emissions through the use of this cleaner fuel, and advanced coal combustion technologies with associated increases in coal efficiencies can be introduced in those cases where coal remains as an economic choice, such as parts of the Urals and Siberia.

As also explained in Chapter 5, it is possible that some 50,000MW of new and refurbished thermal power generating capacity will be required by the year 2010 to replace old equipment and re-equip with more efficient plant if the present industrial decline is reversed, and some 8,500MW of hydro-electricity generating power may also be required. The production of steam and hydro turbines for such a programme would appear to be well within the production capacities of LMZ which has achieved an annual output equivalent to 10,000MW in the past, although allowance would need to be made for a changing product mix including a higher proportion of gas turbines with their associated increased complexity.

At present, however, it appears that the major preoccupations of the electricity power generation industry in the former USSR have been that of privatisation and energy price liberalisation, with expectations that these processes would provide the preconditions to attract investment into the industry. This has not been achieved in practice, however, as fuel prices have risen at a faster pace than those of electricity, and the general economic uncertainties in the former USSR have influenced the direction of private funds in that region away from power utility construction and manufacturing towards trading, where the lead time between investment and income is shorter. Secondly, the most recent financial data available at the commencement of this research showed that a loss of some R800bn ($0.8bn) was incurred in 1993 from some R7553bn ($8.1bn) sales of electricity from thermal power stations (not counting payment of some 8 per cent of sales turnover to a national power investment fund), and a loss of some R150bn ($0.17bn) was incurred from a turnover of some R3186bn ($3.43bn) from heat sales from thermal power stations (again not counting payment of some 8 per cent of sales turnover to a national power investment fund). These losses were offset to a certain extent, however, by an operating profit of some R650bn from hydro-power stations. The financial problems encountered through loss-making activities have been made even more acute by customer non-payment, which accounted for some R7000 billion during 1993. Most of these losses were covered by bank loans which facilitated the use of depreciation allowance for the replacement of some 10,000MW of capacity (or approximately 6 per cent of total non-nuclear capacity) (IEA, 1994, pp. 200-3).

Thirdly, it will be even more difficult for individual power stations to make a profit, as fuel costs increase approximately in line with inflation in the case of gas (which since March 1995 has now reached some 2/3 of the world market price),[3] whilst coal is now priced at international levels.[4] Electricity prices, however, are partly controlled by local political authorities, and do not necessarily increase in step with fuel prices which can account for some 33 per cent of total power generation costs (IEA, 1994, pp.201-2). Although significant increases in electricity prices were allowed in 1993,[5] it is far from clear whether similar increases in electricity prices will be possible in future years; and the prices of domestic heat and power have to account for the budgetary realities of the average citizen during the harsh Russian winters.

It is apparent from discussions in the Russian technical and commercial literature, however, that policy-makers are concerned about the likely future shortfall in power generation capacity in the medium and long term; and the selection of appropriate combustion technologies to meet those shortfalls. Gas is generally regarded as the future dominant fuel for reasons of cost and cleanliness, but there is some uncertainty over the proportions of power generation capacity to be supplied by coal-firing or nuclear fission. Furthermore, although power generation by means of gas turbines and combined cycle systems provide additional advantages of efficiency there is little evidence of domestic investment in such cost effective systems in the former USSR, except for three projects which have also attracted Western investment (see Chapter 6). For coal-fired projects the situation is even more uncertain, with little evidence of any refurbishment of existing power stations or construction of new coal-fired utilities. Rationalisation is taking place amongst the nuclear-powered stations, through various inter-governmental programmes.

The lack of investment in the power generation sector is having a consequent effect upon power engineering equipment supplies, as demand for such equipment has plummeted, and most of the power engineering organizations are large vertically integrated establishments with few alternative markets. The major challenge currently being faced by power engineers in the former USSR is therefore short term survival, and retention of capability for when output may improve in the medium term.

INCENTIVES FOR WESTERN EXPORTERS

Market Opportunities for Power Station Efficiency

Western companies' continued interests in the former Soviet market will depend upon expected short term sales and future strategic advantages. Expected short term sales will also be viewed in the context of current prices, delivery conditions and creditworthiness, whilst long term strategy will be conditioned by such factors as market growth and stability, the level of competition and potential for income from alternative markets, especially those in the Pacific Rim. In addition, long-term strategic decisions will need to include choices about channels for sales, including direct exports, licensing and industrial co-operation, and the protection of confidentiality for the technology transferred.

The present economic conditions prevailing in the former USSR are not conducive to high levels of outlay of Western currency for the purchase of pollution control equipment. This suggests, therefore, that there may only be scope for the marketing of improved power station efficiency procedures, as the capital outlay for such technology can be comparatively small and such outlays can be retrieved in a short space of time by reduced operational expenses from improved efficiencies. Recent investigations by the Finnish Energy Conservation Group, in two combined heat and power stations in the Russian North West (Novgorod No. 20 gas fired, and Petrozavodsk No. 1 coal-fired) demonstrated that improvements in operating and maintenance procedures, and repairs in these power stations, led to substantial savings in fuel consumption and electricity conservation. In the case of the Petrozavodsk No. 2 station, for example, annual savings in fuel oil of 3 GWh per year and electricity conservation of 4GWh per year were achieved through improved combustion control and leak prevention leading to annual savings of $69,000 per year with no additional investments required (Finnish Energy Conservation Group, 1995, pp. 8, 29-32).

In the case of the Novgorod No. 20 station, gas savings of 5GWh per year and fuel savings of 20GWh per year were achieved through improved flue gas analysis, attention to the high pressure and low pressure pre-heaters and improved operation of the feed water pumps. These improvements yielded a saving of $416,000 per year, with an investment of only some $1,000 in instrumentation. Savings in fuel consumption were also obtained from other industrial establishments in North West Russia in district heating stations and pulp and paper mills (Finnish Energy Conservation Group, 1995, pp. 8, 9, 26-8).

In certain cases, though, the required reductions in energy consumption and pollution emissions may require substantial capital investments in new equipment. Further investigations were carried out by the Finnish Energy Conservation Group into the Novgorod No. 20 and the Petrozavodsk No. 1 combined heat and power stations, together with investigations into the St. Petersburg No. 2 station with a view to saving energy through local investments. The investments recommended in these power stations related to such items as insulation and replacement of steam turbine condensers. Total investment of over $5M was recommended for the three power stations with anticipated annual savings of some $4.5M as a consequence of savings of some 323GWh of fuel per year and some 174GWh of electricity conservation. Similar savings were also apparent from district heating systems and pulp and paper mills. The proposed investments would also achieve total reductions of 820 tons per year of SOx from the three power stations and some 190 tons per year of NOx (Finnish Energy Conservation Group, 1995, pp. 8, 26-34, 63)

The Finnish Energy Conservation Group (1995, pp. 68-72) also suggested a number of more substantial projects requiring Western technology mainly for instrumentation and boiler modification, and Western financing. These amounted to some $11.5M in total for the Novgorod No. 20 and the Petrozavodsk No. 1 plant, and were expected to yield annual savings of $3M per year, and emission reductions of 980 tons per year of SOx and 150 tons per year of NOx. In addition, the Group also recommended the installation of a new power plant at St. Petersburg No. 2, which was expected to cost some of $76.4M, but yield an annual saving of some $13.4M.

The report also pointed out the difficulties in obtaining Western financing for these projects, however, and suggested that investigations be carried out into the marketability of products provided as payment-in-kind by industrial customers, and the sales of electricity into the West through interlinking with the Finnish transmission system. Approaches were also suggested to the World Bank's energy efficiency programme for Russia, contracts with special purpose project companies funded by the European Bank for Reconstruction and Development, and multilateral technical assistance funds such as the European Union's Technical assistance for the CIS (TACIS) programme. The Group also highlighted that technical assistance to the energy sector was one of the principal items of technical assistance to Russia, citing an OECD estimate that approximately some $155M (or 10 per cent of total technical assistance) had been allocated to the energy sector between January 1991 and the end of 1993. A substantial proportion of this assistance was for improvements in nuclear safety, but help was also being provided to a large number of energy efficiency projects. The results of these studies need to be interpreted with some caution, however, as a consequence of assumptions of fuel prices and subsequent variations in this parameter together with variations in the exchange rate between the rouble and Western currencies. In the case of the Finnish Energy Conservation Group project, for example, heavy fuel oil was assumed to cost $55 per tonne, natural gas was assumed to cost $77.40 per thousand cubic metres, and electricity was assumed to cost $18.40 per MWh (Finnish Energy Conservation Group, 1995, p. 8).

It is not entirely clear from that report, however, as to whether those costs refer to 1992 when the project was commenced, 1993 when the technical reports and feasibility evaluations were submitted, or 1995 when the final summary report was submitted. The cited natural gas price, for example, differs from a November 1994 retail price of 62,679 roubles (approximately $20) per thousand cubic metres, and 137,809 roubles (almost $40) per thousand cubic metres compared with a European border price of $63 per thousand cubic metres (Stern, 1995, p. 35). The electricity price also differs from a 1993 IEA cited average revenue of 11.6 roubles/kWh (and a cost of some 13.6 roubles/kWh) (IEA, 1994, pp. 200-3) which approximates to some $12.50 per MWh for average revenue and some $14.7 per MWh at cost (assuming an exchange rate of almost 1000 roubles to the dollar at that time). For the end of 1993, however, the IEA reported a price of $38/MWh for industrial customers (IEA, 1994, p. 192). These price variations in both fuels and electricity illustrate the problems of attempting to carry out investment analysis for power stations in the former USSR.

Opportunities for Western Investment in Power Stations

Western investment in the former USSR has been extremely small in aggregate, as illustrated by the EBRD statistic that in 1993 the former centrally planned economies accounted for approximately 10 per cent of the foreign direct investment going into developing countries, and Russia (with 40 per cent of the population and the majority of the land area) received only 21 per cent of this investment ($1.1bn within a total of $5.3bn for the region) in that year. The investment levels for the other former Soviet republics were far lower than this figure. Cumulative data for the years 1990-93 differ for the volume of cumulative foreign investment received by Russia ($2.0bn or $3.15bn) and for the total FDI in the former Soviet and East European region (either $12.4bn or $19.5bn), but in both cases the proportion received by Russia was only some 16 per cent. In terms of committed Western investments during 1990-93, the Russian figure stood at $11.6bn, which was lower than for the Czech Republic ($11.7bn) and only slightly higher than for Hungary ($9.6bn) and Poland ($10.0bn) (Bradshaw, 1995, pp. 8, 9).

The reasons for this situation are partly due to the investment objectives of Western companies and partly to inhibiting factors within the region itself. Although Western companies are aware of a large market potential, and of the importance of the region as a source of materials and components in a global supply strategy, there are concerns over the levels of access to markets and supplies. Furthermore, these concerns are reinforced by uncertainties about political and economic stability to provide the necessary background to achieve consistently acceptable rates of return.

In the case of power generation and power engineering projects, these concerns become even more acute as the key features of such projects are that they are technologically demanding; involve high construction costs; and have operating costs consisting mainly of fuel expenditures (Baragona, 1995). There are therefore attendant technological risks at the construction stage that a power station will not be completed to specification by the required completion date. These risks can be increased by the transfer of Western technology which is unproven in CIS operating conditions, and compounded even further by possible delays in the assimilation of Western technology in CIS power engineering suppliers. Furthermore, these technological risks can be additional to the usual risks of lateness encountered in the timely completion of complex projects, and the former-Soviet region has not previously had a high rate of timely completion of projects as a consequence of harsh weather, inadequacies in the provision of finance throughout a project life cycle, and bureaucratic delays (Amann, 1982). These risks in technology and project management have consequent cost risks as a result of late project completion and delayed revenue earnings, or higher costs to rectify project completion date slippages.

Furthermore, as fuel costs account for a high proportion of total operating costs, fuel supplies of adequate quality must be guaranteed over the life of the project, and the prices charged for these fuels must be transparent in terms of available supplies and other alternatives. In some regions of the former USSR, guaranteed supplies of fuel are not always secure, except for gas in the Russian Federation and Kazakhstan and some other regions of central Asia; and coal in East and West Siberia, Kazakhstan and parts of Central Asia and Ukraine.

In addition, a market of sufficient size and duration must exist to generate the requisite streams of income to cover operating costs, provide an adequate profit to the project company and repay the required loans to the project sponsors. The pricing system for electricity in the former USSR is not readily transparent to Western companies as the regional governments continue to have some influence on the prices charged. Finally, the development of the legal and taxation frameworks are far from complete or transparent in the former USSR to enable business to be conducted on market economy principles and practices, which makes Western project financiers even more cautious about additional commercial risks to be encountered in the market (Petkovich, 1995; Mainelli, 1995).

The risks of project finance in the former USSR, have to be balanced against investors' perceptions of opportunities in other markets. It is apparent, therefore, that Western investment in the power generation industry in the former USSR is likely to minimise risk as much as possible, and this strategy is evident in the reports of companies active in the region. The income from the new 'North West' combined cycle power station, for example, is partly to be covered by electricity sales to Finland, whilst the *Nizhnyi Novgorod* and *Mostovski* power generation prospects are to be funded through the sale of materials to Western markets.

Opportunities for Western Investment in Power Engineering

Major Western investors in power engineering projects in the former USSR have high levels of expertise in the sale of commodities available from the former Soviet region, on Western markets (see Chapter 6). A recent report of investment in a Kazakh power station *(Uralsk)* by a Belgian power generation facility, for example, also refers to the purchase of Kazakh uranium for use in Belgian nuclear power stations by another division of the Belgian utility. The report does not explicitly link these two commercial transactions, but the supply of the Kazakh uranium could also possibly be used as a source of finance for the 500MW co-generation project, estimated to cost some $500M (Tractebel, 1995).

In view of the severe shortages of foreign currency in the former USSR, and other priorities for its use, there will probably be little funding available for the import of Western pollution control equipment, although finance at competitive rates may be available from multilateral sources of funding. Western companies will consequently have to consider various forms of technology transfer such as licensing and industrial co-operation in order to achieve a commercial transaction, whilst simultaneously using materials and labour which can be resourced more cheaply.

Companies taking a long term view of the market may also consider factories and design organisations within the former USSR as cost-effective sources of supply to service other international markets on a global scale. Such sourcing may be particularly attractive for components having a high labour content, or where the volumes of Western output are beginning to fall as a consequence of obsolescence, although a significant demand may also continue for components and sub-assemblies as spares.   In the case of industrial gas turbine production, where the majority of the world market is shared between ABB, Siemens KWU, MHI, GE, GEC-ALSTHOM and Rolls Royce, access to cheaper production costs could provide signficant international competitive advantage.

In addition, highly trained Russian and East European designers can be recruited for a fraction of the cost of their counterparts in Western Europe and USA, and the use of CAD systems within an international electronic data interchange (EDI) framework provides an opportunity for the reduction of technical development costs.   These long term strategic options will probably require high levels of investment from Western companies, particularly in terms of quality management and supply chain traceability procedures in the case of manufacturing, and CAD systems in the case of design.   Such costs may act as a disincentive for many companies, particularly in the present climate of political and economic uncertainty in the former USSR.   A positive approach from a few large Western companies, however, will probably have the effect of attracting other organizations to adopt a defensive competitive strategy in that market, and provide investment to achieve that strategy.

CONCLUSIONS

This chapter has highlighted the incentives and disincentives to the development of low SOx and low NOx power generation facilities in the former USSR, from both Eastern and Western viewpoints. This chapter demonstrates that although the technical preconditions have been shown to exist for a transfer of Western technology to the former USSR (see Chapter 6), and trade opportunities exist for the sale of this equipment and know-how, there are various business barriers which prevent these transfers from taking place. These include general uncertainties about political and economic stability in the former USSR, or shortage of investment capital within the region itself, and specific concerns relating to fuel prices, project management and market transparency.

Until these concerns are lessened, therefore, it is likely that technology transfer will only take place on a relatively small scale, and that the majority of this technology will relate to gas-fired combustion technology rather than coal. Gas-fired systems can normally be constructed in a shorter time interval than can their coal-fired counterparts and can also achieve high efficiencies of operation. The problem is, however, that emission reductions from coal-fired systems are more urgent from the environmental viewpoint particularly in Siberia, Ukraine and Kazakhstan; and the development of further 'clean' coal-fired systems could also act as an incentive to further development in the coal-mining industry in those regions. Such progress, however, will probably require further investment support from multilateral agencies in view of the risks and timescales involved; and further design and manufacturing development work across a range of fuels by fuel and boilermaking companies in both the West and the former USSR.

NOTES

1. Calculated from the following data provided by OECD, Foreign Trade by Commodities, Series C, 1992 ($M).

Year	1987	1988	1989	1991	1991	1992
Total OECD Exports	47,678	56,199	60,956	70,957	73,637	80,237
OECD Exports to Canada	3,354	3,903	3,893	4,960	5,218	5,526
OECD Exports to USA	9,175	10,988	12,004	13,599	13,230	14,793
OECD Exports to EC	16,826	20,692	23,569	27,514	28,614	31,037
OECD Exports to Far East	5,209	5,792	7,062	8,828	9,767	10,701
OECD Exports to China	602	683	897	1,244	1,005	1,399
OECD Exports to former USSR	233	211	223	256	369	322

2. The $6-8bn estimate is for the improvement of existing Polish power plants to an 'A' or 'B' standard, with new power plants being built to a 'C' standard IEA, (1994), p. 160. The estimate of $20bn over 15 years is provided by Oudhuis et al., (1993).
3. In March 1995, industrial retail gas prices were increased from some 62,679 R/mcm to some 137,809R/mcm, and excise taxes were also increased to 25 per cent of the total price. This figure converts to $39.20/mcm using the current exchange rate, compared with a European border price of some $63/mcm (not counting transport costs to the Russian border) (Stern, 1995).
4. Government regulated gas prices are maintained below the level of free market prices for steam coals on a BTU basis. High rates for railway transport of coal exceed free market rates by some 35-40 per cent (see Zykov, 1995).
5. Electricity prices for industrial consumers increased from some 4.15 R/kWh to 8.07 R/kWh during the first two quarters of 1993, and then jumped to some 42.06 R/kWh during the third quarter of that year and 45.39 R/kWh during the fourth quarter. For residential prices, the levels for the same time periods were some 0.24 R/kWh, 0.36 R/kWh, and 6.00 R/kWh for both the third and fourth quarters. For the third quarter of 1993, the dollar equivalents of the Russian prices were some $0.038/kWh for industrial users, and some $0.005/kWh for residential prices. (IEA, 1994, p. 192).

REFERENCES

Amann, R. (1982), 'Industrial Innovation in the Soviet Union, Methodological perspectives and conclusions' in Amann, R. and Cooper, J.M. (eds), *Industrial Innovation in the Soviet Union*, New Haven, CT: Yale University Press, pp. 1-38.

Baragona, K. (1995), 'Legal Considerations and Structure of Contracts in Financing Power Projects', paper presented at Power-Gen International seminar, 15 May 1995, Amsterdam.

Bradshaw, M. (1995), *Regional Patterns of Foreign Investment in Russia*, London: Royal Institute of International Affairs.

Boehmer-Christiansen, S. and Skea, J. (1991), *Acid Politics: Environmental and Energy Policies in Britain and Germany*, London: Belhaven.

Investing for a better environment, EBRD (undated).

Economic Combustion for Europe (ECE) (1994), Executive Body for the Convention on Long Range Transboundary Air Pollution (LRTAP), *1994 Major Review on Strategies and Policies for Air Pollution Abatement: Tables and Figures* (EB.AIR/R.87/Add.1) 21 September 1994.

Financial Times, 28 October 1993, p. 38 ('Russia and EC agree outlines of solution to aluminium export row').

Financial Times, 6 January 1994 ('Collateral plan may solve CIS Aluminium Problems').

Financial Times, 13 January 1994 ('Aluminium flood unites US and EU').

Finnish Energy Conservation Group (1995), *Energy Conservation Study of Nine Industrial and Energy Ulility Plants in the Russian Federation:*

Final Summary Report, Ministry of Trade and Industry, Finland, Studies and Reports 87/1995.

Fraser, P. (1992), *The Post-Soviet States and the European Community,* London: Royal Institute of International Affairs.

Goskomstat (1988), *Narodnoe Khozyaistvo SSSR v 1987g: statisticheskii ezhegodnik,* Moscow: Finansy i statistika.

Goskomstat (1990), *Narodnoe Khozyaistvo SSSR v 1989g: statisticheskii ezhegodnik,* Moscow: Finansy i statistika.

Hewett, E. (1984), *Energy, Economics and Foreign Policy in the Soviet Union,* Washington, DC: Brookings Institution.

Hiltunen, H. (1994), *Finland and Environmental Problems in Russia and Estonia,* Helsinki: The Finnish Institute of International Affairs, (Foreign Challenges, 1994).

International Energy Agency (IEA) (1988) *Emission Controls in Electricity Generation and Industry,* Paris: OECD.

International Energy Agency (IEA), (1994), *Electricity in European Economies in Transition,* Paris: OECD.

International Monetary Fund (IMF), the World Bank, Organization for Economic Co-operation and Development (OECD), the European Bank for Reconstruction and Development (EBRD) (1991), *A Study of the Soviet Economy, Volume 3,* Paris: OECD.

Mainelli, A.L. (1995), 'Financing Power Projects in Central/Eastern Europe: A Developer's Perspective' Paper presented at Power-Gen International Seminar, 15 May 1995, Amsterdam.

Lyalik, G. N. and Reznikovskii A.Sh. (1995), *Elektroenergetika i priroda,* Moscow: Energoatomizdat.

OECD (1994), *Foreign Trade by Commodities, Series C, 1992,* Paris: OECD.

Oudhuis, A.B.J., Jansen, D., Iwanski, Z. and Golec, T.W. (1993), 'Repowering of Polish Coal-fired Plants with IGCC', Presented at the Twelfth EPRI Conference on Gasification Power Plants, 27-29 October 1993, San Francisco, CA: Hyatt Regency.

Petkovich, D. (1995), 'Financing Mechanisms for Coal Projects' Paper presented at the EC/Directorate for Energy/DG XVII/Rosugol Conference on 'Coal in a Competitive Market', Academy of Management, Moscow, 21-23 June 1995.

Rosengaus, J. (1986), 'The Selection of Steam Parameters for Soviet Thermal Power Plants', in Young, K.E. (ed.), *Decision Making in the Soviet Energy Industry: Selected Papers with Analysis,* Delphic Associates, Falls Church, VA, pp. 1-31.

Rutland, P. (1992), *Business Elites and Russian Economic Policy*, London: Royal Institute of International Affairs.

Sakwa, R. (1993), *Russian Politics and Society*, London: Routledge.

Stern, J.P. (1995), *The Russian Natural Gas 'Bubble': Consequences for European Gas Markets*, London: The Royal Institute of International Affairs, Energy and Environmental Programme.

TACIS Contract Information Budget (1993) (European Commission, January 1994).

TACIS Contract Information Budget (1994) (European Commission Part I May 1994 and Part II July 1994).

Tractebel (1995), 'Tractebel is awarded a cogeneration project in Uralsk (Kazakhstan)', Tractebel Press Release, 1 December 1995.

Tuovinen, J.P., Barrett, K. and Styve, H. (1994), *Transboundary Acidifying Pollution in Europe: Calculated fields and budgets 1985-93*, Det Norske Meteorologiske Institut Technical Report No.129 (EMEP/MSC-W Report 1/94), Oslo, July 1994.

Whitefield, S. (1993), *Industrial Power and the Soviet State*, Oxford: Clarendon Press.

Zykov, V.M. (1995), 'Basic Factors for the Improvement of the Competitiveness of Promising Russian Coals', Paper presented at the EC/Directorate for Energy/DG XVII/Rosugol Conference on 'Coal in a Competitive Market' 21-23 June 1995, Academy of Management, Moscow.

8. Comments, Conclusions, and Suggestions for Further Research

COMMENTS

Scope of the Research

The research described in this book has focused on one particular aspect of atmospheric pollution in the former USSR, namely the emissions of oxides of sulphur and nitrogen from the electricity generation sector. These topics were chosen because of the important effects of SOx and NOx pollutants in the production of acid rain, and the levels of cross-boundary transfer of these acidic compounds to other countries. The power generation sector was selected for particular study, as this industry is accredited to be the largest single industrial emitter of SOx (43 per cent) and NOx (59 per cent) pollutants.

As pollution control has been a major field of technical development in Western countries to meet recent legislation on power station emissions, proven Western technologies exist to resolve some of the SOx and NOx emission problems currently being faced in the former Soviet region. Furthermore, as increased efficiency continues to be demanded from modern power generation machinery for Western utilities to be competitive, contemporary Western technologies provide the means to simultaneously improve power station performance whilst reducing harmful atmospheric emissions.

The major constraints on the transfer of these technologies to the former USSR, however, are the suitability of Western technologies for available fuels and operating conditions in the former Soviet region; the technical capabilities of the region's power engineering and electricity generation facilities to assimilate the requisite know-how; and the financial constraints on investment in that region as a result of the recent political and economic instabilities. The research described in this book has consequently addressed the issues of technology transfer within the context of industrial realities in the former Soviet region, in order to identify the options available to achieve reduced atmospheric pollution in the former USSR and the potential for Western technology to assist in that process.

Research Methods

The research methods used during this study can be categorised into four main groups, namely:

1. A review of Western literature related to the topic, including industrial reports on equipment design and performance;
2. A review of material in the Russian language paying particular attention to technical and industrial sources in this field (for example, *Energetik, Elektricheskie stantsii, Teploenergetika, Tyazheloe mashinostroenie, Energo-mashinostroenie);*
3. On-site structured interviews of Western companies having proven expertise in SOx and NOx reduction technologies, and experience of trade and technology transfer to the former USSR. The thirteen Western organisations selected for interview were located in the UK, Germany and Finland;
4. On-site structured interviews with Russian companies with proven expertise in the power engineering industry. All of these interviews took place during the spring of 1995 and 1996 at four organisations located in St. Petersburg.

At the commencement of this research, it was intended to survey Russian power engineering establishments in a number of locations, particularly turbine manufacturers and boiler development organisations in St Petersburg, Moscow or Taganrog. Applications for discussions at boiler factories in Moscow and Taganrog met with no response, however, but interviews were obtained at three establishments in St Petersburg *(Leningradskii Metallicheskii Zavod [LMZ], Leningradskii Zavod Turbinnykh Lopatok [LZTL], Tsentral'nii Kotel'no-Turbinnyi Institut im. I.I. Polzunova [TsKTI]).* These interviews were made possible through contacts between the author's University Department and the St Petersburg State Engineering Economics Academy, contacts between that Academy and a St. Petersburg power engineering service company *(Energiya-Servis)* and contacts between that service company and other power engineering establishments in St Petersburg city.

Many Russian factories have received large numbers of visiting Western delegations with only limited business opportunities being created from these visits, and Russian senior management are consequently questioning the economic benefits to be gained from this activity. This scepticism has become even more apparent in relation to research, which may not provide short-term income benefits to assist in the quest for survival, nor necessarily guarantee medium- to long-term product innovation or efficiency advantages. In many ways, therefore, although individual Russian managers are usually more accessible than in the former Soviet period and more open in their discussions, the pressures on time and financial resources do not always allow the possibilities for research to be translated into realities. It is apparent therefore, that access for research into Russian enterprises still requires assistance and support from intermediaries, as well as persistence from the Western researcher. Without the prior existence of these academic and industrial contacts, the scope of the case study component of this research would have been severely limited; but the continued collaboration with these contacts has provided the access for longitudinal study as the former USSR continues to undergo major political and economic change.

It was also anticipated at the commencement of this research that detailed data would be obtained on process planning specifications to compare with Western tolerances. Available data on the details of these specifications were limited, however, as they are now regarded as proprietary information, and the growing commercialisation in the former USSR has further developed the tendency to protect all types of proprietary data unless some economic reward is available for its provision. Furthermore, the present author was unable to locate any published data on typical tolerances observed in the power engineering industry, similar to that previously existing for other types of engineering production (Hill and McKay, 1988; Hill, 1991). Russian technical capability has therefore been assessed by the capability of factories to meet the requirements of Western customers or Western partners, as viewed by those partners; or to meet the requirements of Western quality control organisations such as TUV. Continued contact with Western companies having interests in manufacturing in the former USSR, however, may provide further data on Russian process planning specifications.

CONCLUSIONS

Volumes of Emissions

Data from various published sources were used to obtain estimates for the volumes of SOx and NOx emissions from the former USSR in general and the power generation sector in particular, for the base years of 1980 (for SOx) and 1987 (for NOx), and selected time intervals beyond those dates. Data from these various sources and its subsequent analysis using a range of assumptions, revealed wide variations in the estimates of both SOx and NOx emissions. The process of estimation was made even more difficult by the tendency of the published sources to refer to a wide variety of geographical regions, namely either European Russia, the Russian Federation, European regions of the former USSR, or the USSR in total.

It was consequently only possible to provide estimates for a range of emissions, and these varied between 16.2 and 28.5M tonnes of SOx for the former USSR in 1980, and between 4.5 and 11.2 Mtonnes of NOx for the former USSR in 1987. For the electricity generation industry, the emissions were estimated to vary between 7.0M and 10.6M tonnes of SOx in 1980 and to approximate to 2.7M tonnes of NOx in 1987 (see Chapter 2).

The estimates of total SOx and NOx emissions from the former USSR were found to be far higher than those of UK and Germany during the base years of 1980 and 1987 (5.3M tonnes of SOx emissions for UK and 3.6M tonnes for Germany; 2.6Mtonnes of NOx for UK and 3.1M tonnes for Germany). Estimates of SOx and NOx emissions from the former-Soviet electricity generation sector were also higher than for the UK and Germany (3.8M tonnes [UK] and 2.2M tonnes [Germany] of SOx in 1980, and 0.8M tonnes [UK and Germany] of NOx in 1987).

The levels of aggregate SOx emissions from the former USSR were therefore nearer to those of the United States for SOx emissions (23.8M tonnes in 1980) than to West European countries, although the levels of Soviet NOx emissions were found to be far lower than the American figure (18.6M tonnes in 1987). The levels of officially reported SOx emissions reflect the levels of electricity generated in both of these countries (2,094TWh for the USA in 1980 and 1,294TWh for the former USSR), although the USA would appear to be using cleaner coal combustion technologies if allowances are made for the possible under-estimate of the Soviet figure, and the comparatively higher proportion of the use of coal in America.[1] In the case of NOx emissions, however, the differences between the two countries can be explained by the higher levels of road transport in the USA.

Fuels Policy

A major factor influencing SOx and NOx emissions on a national level is the selection of fuels in the total combustion mix, particularly for power and heat generation. The data presented in this book (see Chapter 3) have shown a continuing trend in the use of natural gas for power generation throughout the former USSR. This trend has been driven by the continued difficulties in the extraction of oil combined with an established international market for that commodity; and a reduction in the calorific value of coal from existing mines.

The combustion of solid fuels (mainly Kansk Achinsk and Kuzbass coals) is important for heat and electricity production in the regions of Western and Eastern Siberia, the Urals, and Ukraine. From information presented in this research however, there is little evidence to suggest that aggregate levels of emissions from the former USSR reduce coincidentally or proportionately with the volume of electricity produced in that region. On the other hand, as other information in this research has shown that most coal-fired power stations are too old to justify high levels of investment in emission control equipment, levels of emissions are likely to increase if industrial output moves out of recession. Oil will remain as an important fuel in some regions of North West Russia, and as a reserve fuel in other power stations. As a consequence of continual change in the fuel mix, therefore, it appears that SOx and NOx emissions will become less of an aggregated problem for the former Soviet region as a whole, and more of a local region where solid fuels and heavy fuel oils continue to be used.

Technological Options

The research described in this book has provided information on the wide range of technological options for SOx and NOx emissions available for transfer from the West to the former USSR, paying particular attention to existing technological expertise in the former Soviet region (see Chapter 4). The particular technologies considered as the most cost effective candidates for transfer include low NOx burners for gas-fired and oil-fired boilers (an area of technology in which Russian technologists have been shown to possess existing technological expertise), high efficiency gas turbines for use individually or as major components of combined cycle systems, gas engines for use in very small power stations in remote areas, low NOx burners for new and existing coal-fired power stations, and fluidised bed combustion for new coal-fired installations.

Flue gas desulphurisation (FGD) systems have also been proven for the reduction of SOx emissions from solid fuels, but the cost of these systems may prove too expensive for installation in existing power plants because of their age. It may be a viable option to install FGD units in new power stations, however, where a proportion of the high initial costs may be recouped through simultaneous introduction of improved working practices. At the present time, however, investment in new coal-fired power stations appears to be a very remote possibility in the former USSR in view of the anticipated cost, the existence of excess power generation capacity in the former USSR in the short to medium term, and a continuing preference within the Russian Federation for the more widespread use of natural gas.

In view of the anticipated important role to be played by coal in the long term, particularly in Siberia and Ukraine because of the comparatively high levels of coal reserves in those regions and a shortage of gas in Ukraine, it is apparent that development and pilot production should be commenced for the implementation of 'clean' combustion technologies using solid fuels from the former Soviet region. These technologies include fluidised bed combustion which enables SOx emissions to be reduced through the introduction of calcium compounds, and lower NOx emissions to be achieved as a consequence of lower combustion temperatures; and integrated gasification combined cycle systems (IGCC), in which coal is gasified to provide fuel for a gas turbine, and also as auxiliary heating in a heat recovery steam generator.

There is still significant development work to be done on these systems, however, in view of current operational problems associated with vibration and corrosion. It is important to note the long term importance of such development work in clean coal technologies, even though these developments may take place in Siberian regions which are a long distance from Europe with its current environmental concerns of cross-boundary pollutant transfer from the former USSR. In addition, there are highly trained combustion engineers and scientists in the former USSR with expertise in these fields of work, who could play a significant role in product and process development.

Hydropower also remains as a potential source of electricity generation in the former USSR, and *LMZ* is a well established manufacturer of hydroturbines (see Chapter 6). 'Low-head' helical turbines (Gorlov 1995) may emerge as potential sources of hydrogeneration, if the scope for larger systems is limited.

Power Generation in the former USSR

This research has analysed available data on power station capacity in the former USSR, paying particular attention to the individual capacities of each power station, the types of fuel used, and the age of the power station for large utilities (see Chapter 5). From these data, it has been possible to estimate the distribution of capacities of power stations according to fuel type and geographical region; and to identify candidate power stations for investment in pollution reduction equipment. In addition, estimates are provided on future power generation capacity (some 40,000-50,000MW of new or refurbished capacity by 2010 emerging as a consensus figure amongst Russian commentators), and the fuel requirements necessary to fire this capacity.

Western Companies and Technology Transfer

All of the major Western power engineering companies are active in the former USSR, but particularly in Russia; although the level of activity varies from company to company (see Chapter 6). The leading actor in power gas turbine technology in the region would appear to be Siemens Kraftwerkunion, which has a joint venture arrangement with the largest civilian power turbine manufacturer in Russia (*LMZ*, St Petersburg) and installation programmes in three major power stations in St. Petersburg, Nizhnyi Novgorod, and Krasnodar. ABB has also developed gas turbine power engineering facilities at the *Nevsky Zavod* factory (St Petersburg), and GEC-ALSTHOM and GE are developing production arrangements with the Kirov factory in the same city. These latter facilities produce turbines of lower capacity than the larger machines manufactured by *LMZ*, but the ABB, GEC-ALSTHOM and GE machines are capable of being used in smaller power stations of which there are a large number in the former USSR (although *LMZ* can also supply 40MW and 60MW power generation turbines), and as gas pumping drive equipment. Furthermore, ABB are investing significant sums of capital in Poland and the Czech republic to provide factories which serve the international market.

These West European and American companies have therefore achieved a commanding position in the supply of gas turbine equipment in the former USSR, and can justify the frequently used statement of 'claiming the market' in anticipation of future development. The position of British and French companies in the power generation gas turbine market would appear to be less advanced than those of their Western European competitors, however, as the majority of effort has been devoted to aero-engine and gas-pumping applications. This position may change, though, through the sales potential for smaller capacity GEC-ALSTHOM gas turbine products and strategic alliances between Rolls Royce, Westinghouse, Mitsubishi Heavy Industries and Fiat. There is also a potential for using Rolls Royce aero-engine derivatives as gas-generators to provide a hot gas stream to drive a power generation turbine, or in conventional turbo generator configurations. At the time of writing, however, one of the companies in the Rolls Royce Industrial Power Group (Parsons Power Generation Systems has recently been sold to Siemens Kraftwerkunion (see Chapter 6) and another (International Combustion Limited) is for sale, and the position of these companies in the Russian market is therefore less certain than when this research was commenced.

High levels of competitiveness are also evident in solid fuels combustion, with all the major West European competitors being active in the former Soviet region, although the level of activity is lower than for gas-fired systems. It is apparent, therefore, that further developments, probably aided by unilateral and multilateral funding could be used to assist in these programmes and reduce some of the perceived disadvantages of location (Siberia and the Russian Far East) and project risk. Useful development programmes have commenced in clean coal combustion (low NOx burners and CFB systems) in which a range of firms are active.

Interviews with Western companies did not provide detailed process specification data or details on manufacturing audit procedures, but a general respect for Russian production capability was evident. Improvements will be required, however, in manufacturing processes for aero-derivative turbine components, traceability procedures and workplace quality management, to meet the quality assurance requirements of Western markets. It is apparent that Western investment in these activities is likely to continue and should be monitored, and closer contact established with companies carrying out a manufacturing audit in that region. In addition, studies should be commenced of the investment programmes of Japanese power engineering companies in the former USSR, as this present research has been concerned mainly with power engineering companies from Western Europe and USA; although one UK company included in this study (Babcock Energy) has recently been acquired by a Japanese conglomerate (Mitsui Engineering and Shipbuilding) (see Chapter 6).

Power Engineering in the former USSR

A study of a sample of power engineering companies in the St. Petersburg region has revealed their capability to assimilate many of the Western technologies available for the reduction of atmospheric pollution (see Chapter 6). Particular attention has been paid to gas turbine technology in view of the opportunities available for reduced emissions of SOx and NOx from the use of these machines, providing the combined effects of 'clean' fuels, 'clean' combustion, and increased efficiency.

The data available from the small sample of enterprises studied in this present research suggests a high level of capability in the former USSR in the manufacture of steam turbine components and their assembly into finished products, and similar high levels of capability to assemble and package advanced power gas turbine component manufacturing technology. Heat recovery steam generators for use in combined cycles, and advanced 'clean coal' combustion techniques such as pressurised fluidised beds and integrated gasification appear to be at the development stage rather than in full scale production, however. Furthermore, there is scope for developing workplace based quality management systems in addition to the more usual pass/fail systems of quality control for all types of power engineering equipment.

One particular and major problem being faced by power engineering factories in the former USSR is that of business survival, in view of falling demand and lack of investment in the defence sector which was previously an important market for civilian engineering. Furthermore many civilian customer organisations are failing to meet debt obligations, and non-payment has become almost endemic in the former Soviet economy. In addition, there are significant organisational uncertainties caused by privatisation and the evolving strategies of large holding companies emerging on a national industrial scale, together with concerns over the downsizing objectives of some Western investors. The opportunities for competition between individual civilian power engineering companies have therefore become constrained by the necessity of meeting the objectives of large holding companies; and the major competitor to the largest civilian power engineering group is now viewed as a similar conglomerate supervised by the defence industry.

International Political and Economic Factors

Although the aggregate levels of SOx and NOx emissions have been high from the former USSR, particularly when compared with other European countries, it is apparent that the specific emissions of SOx and NOx per unit of electricity produced were generally lower for the former USSR compared with UK and Germany (5.4-8.2g/kWh for the former USSR, 15.2g/kWh for UK and 7.3g/kWh for Germany for SOx in 1980; 1.6g/kWh for former USSR, 3.1g/kWh for UK and 2.3g/kWh for Germany for NOx in 1987) (see Chapter 7). These data suggest, therefore, that policies for the reduction of SOx and NOx emissions could be based on the implementation of procedures for increased energy efficiency to reduce electricity production levels and consequent quantities of fuel combusted, in addition to the selection of 'cleaner' fuels in the combustion mix and the introduction of advanced combustion technologies.

There is published evidence to suggest that the Russian Federation and Ukraine have met the Helsinki Protocol for SOx emissions, and were also on target to meet the Sofia protocol for NOx emissions, as defined by the United Nations Economic Commission for Europe (ECE) Long Range Transboundary Air Pollution (LRTAP) Conventions. These protocols called for a 30 per cent reduction of SOx emissions by 1993 compared with 1980, and to hold the 1993 NOx emissions at the 1987 levels. although their achievement by Russia and Ukraine have probably been influenced by industrial decline and consequent reduced demand for electricity and fuel; and an increase in the proportion of gas used in the fuel mix to generate electricity.   The LRTAP convention calls for further decreases in atmospheric pollutant emissions and it remains to be seen if these can be achieved if Russian and Ukrainian industrial production begins to increase. Major political and economic pressures may be exerted through the European Union's Large Combustion Plant (LCP) Directive, which is focused towards cleanliness of individual power stations, particularly if producers in the former Soviet Union wish to sell large quantities of electricity into the European Union, or products where electricity accounts for a high proportion of production cost.

The sustained increase in the use of natural gas has led to a consequent reduction in the emissions of SOx and NOx, which has probably been particularly significant in the regions of European Russia, where gas is by far the most dominant fuel because of the high transport costs over long distances for solid fuels. To a certain extent, therefore, the continued use of natural gas in European Russia will have led to a reduction of SOx and NOx emissions in some of those regions of North West Russia where pollutants may be transferred to Western Europe, particularly Scandinavia. There still remain areas of particular local and international concern, however, because of high levels of SOx and NOx emissions from some of the newly independent states which formerly made up the European region of the former USSR (for example, some coal-fired power stations burning high sulphur coals in Ukraine, and oil-shale fired power stations in Estonia), and metallurgical installations and paper mills in North West Russia, and coal-fired power stations and industrial installations in Siberia. Fuel and combustion technology options therefore assume particular importance in those regions of the former Soviet Union.

Commercial Aspects

The research carried out in this project has demonstrated a general reluctance by Western private financial institutions to advance low interest loans for projects related to environmental improvements and power generation in the former USSR, because of perceived high levels of political and economic risk in the region and a shortage of funds in the former USSR itself for initial investment in power generation and environmental projects. Furthermore, other world regions, such as South East Asia and China are assessed as being more stable and providing less risky returns than those available from the former USSR.

It is apparent, therefore, that although many of the technical preconditions are in place for the transfer of Western technology for reduced environmental pollution (for example, proven products, product and process knowledge, technical competences), the commercial conditions for such transfers are far from satisfactory. Exceptions to this general observation include the particular cases where foreign currency can be obtained through re-export of electricity back to the West as in the case of the North West (*Severno-Zapadnaya*) power station in St Petersburg which can be used to generate electric power for export back to Finland. Alternatively, regional government investment in a power station can facilitate the export of a range of products from the region with market potential in the West, or increased efficiencies of generation can provide opportunities for the sale of surplus payments in kind of industrial materials from customers. These in turn, require countertrade expertise by Western suppliers, to define Russian products with Western sales potential and a level of political support from the relevant regional government to permit the export of these products.

It is apparent from the survey of industrial companies carried out in this research that the transfer of gas-fired power generation technology through gas turbine and heat recovery steam generator expertise is occurring at a faster rate than solid fuel combustion technology.   There can be various reasons for these differences in the pace of technology transfer, including a preference for natural gas as the dominant fuel in European Russia. Technological factors, however, also play a major role, including the possibility of transferring gas combustion technology in modular form through gas turbine sets, and the subsequent phased and modular transfer of technology related to particular major assemblies such as rotors and stators, and associated components such as blades.   In the case of solid fuel combustion, however, it is necessary to transfer a range of fabrication process technologies under construction site conditions as well as factory production; and consistencies in product, quality and scheduling can therefore be more difficult to achieve.

Furthermore, the time scale for solid fuel construction projects is often very lengthy, causing even more uncertainties for the providers of project finance. In addition, the environmental benefits available from gas-fired projects are more readily visible to countries adjoining European Russia, whereas environmental benefits from investment in clean coal combustion in the Urals and Siberia are less easily seen in Europe in view of the long distances between the two continents.

It is apparent, therefore, that there is scope for further provision of Western project finance to achieve higher power generation efficiency in the former Soviet region, although private finance may be directed more towards gas-fired projects because of the comparative ease of transferring these technologies, and the expertise of Western companies in the trading of the gas which may be saved through use of more efficient systems.   For solid fuel power stations, multilateral finance will probably be necessary because of the longer construction periods and higher perceived levels of technological and organisational risk with these projects.   In the longer term, however, this situation may change if coal becomes more of an accepted fuel in view of the large reserves, and cleaner coal technologies become more accepted and proven.

Domestic Political and Economic Stability

Since the mid-1980s, it has become increasingly difficult to predict the political and economic directions of the former USSR, and the associated pace of change. The scope of anticipated change to be implemented through

Gorbachev's programmes of *glasnost, perestroika,* and *demokratizatsiya* was almost entirely unexpected, although several Western observers had proposed various strategies by which the former USSR may have achieved improved levels of industrial performance. Furthermore, the fragmentation of the former USSR in 1991 was as rapid as it was surprising, as was the siege of the Russian parliamentary building in 1993 followed by elections with some unexpected results. Tensions have continued between Parliament and President in the Russian Federation, and these tensions have influenced the pace and direction of economic reform and industrial restructuring. At the time of writing, the political vacuums created by the health problems of President Yeltsin remain a cause of concern in Western countries.

In essence, however, although democracy has taken deep roots in the former USSR following the cessation of single party rule, the emergence of major political parties has yet to occur for the development of Parliamentary and Presidential policies. Furthermore, there is always the concern that a desperate electorate may elect a President proposing extreme nationalistic policies, as a reaction to the fall in living standards caused by high inflation associated with the economic policies of democratic politicians.

These political uncertainties are also reflected in economic uncertainties, particularly over rates of inflation and international exchange. Furthermore, the legal framework for incoming investment is far from clear for most of the region, and the procedures are not always apparent, consistent or enforceable. The programme of privatisation has done little to break up the monopoly power enjoyed by many of the large factories, as shortages and inflation precluded the establishment of 'normal' market conditions, and the consequences of these economic factors on the standard of living had caused much of the population to cash their privatisation vouchers. These sales have consolidated the position of existing senior management, holding corporations, and some investment funds of dubious probity. Shortages and inflation have combined to create a recession in industrial output, and an environment in which postponed payment of bills is the norm rather than the exception. Recession in demand has also had a consequential effect on industrial employment, with many skilled and qualified personnel retaining their employment for reasons for access to welfare provision, but only working on a part-time basis because of low levels of demand.

The process of industrial restructuring is consequently very uncertain under such conditions, and it is difficult to predict the outcome of this complex process. Many of the large vertically integrated plants remain in existence, and in the present conditions of supply shortages can guarantee some stability in the provision of components as was their objective in the previous planned economy. This role has grown in importance in some cases following the fragementation of the former USSR with previously important suppliers now being located in foreign countries. Furthermore, many large assembly plants have purchased shares in component supplying factories in order to guarantee sources of supply to meet delivery requirements.

The decline in demand, however, has caused massive increases in over-capacity, which have been partly met by short-time working and a retention of plant and machinery in anticipation of an upturn in demand. This macro-economic environment has influenced the power engineering industry as much as any other, with most factories retaining their physical assets in anticipation of subsequent increases in demand. It is difficult to predict the time interval over which these assets can be retained, however, and the consequent long-term viability of some of the prospective partners for joint ventures in the former USSR. Furthermore, those partners which manage to remain viable may be equipped with an ageing stock of manufacturing machinery.

It is therefore difficult for Western investors to predict satisfactory rates of return from potential investments in power generation and power engineering facilities in the former USSR, and the levels of investment have therefore been chiefly related to defensive strategies to retain existing markets and areas of influence, and to be in a position to 'claim the market' if conditions improve and there is progress out of recession. Most investment has been directed towards creating an infrastructure for improved quality of production, in order to be able to use partners in the former USSR as a cost-effective source of supply to meet Western markets. Strategies of downsizing proposed by Western investors have frequently met with opposition from the Russian partner, however, and it therefore still remains to be seen how Western companies will continue to adapt to commercial conditions in the former USSR. These conditions continue to be influenced by high levels of political instability, high inflation and industrial recession, although some optimism still remains for an upturn in demand, a large market to be satisfied, and a cost-effective location for delivery to Western markets.

Until this optimism is translated into reality, however, there is little likelihood of significantly higher levels of either Western or Russian investment in either power generation or power engineering to achieve substantially lower levels of SOx and NOx emissions consistent with higher levels of industrial output compared with the presently lower levels caused by industrial recession. Furthermore, additional investment by Western companies will be contingent upon increased transparency and consistency of commercial legislation and practice in the former USSR, and a movement towards practices more closely related to those encountered in the West. Changes in these directions may cause increased tensions in the former USSR itself, however, if it is considered amongst the population as a whole that the country is losing national control over its natural resources and assets.

As a final conclusion, therefore, this research has demonstrated that there is a national problem to be solved of high levels of SOx and NOx emissions, and that many of these problems occur in particular industrial locations. The research has also shown that technologies exist in both the former USSR and the West for the reduction of these levels of atmospheric emissions from the power generation sector, and that many of the Western developments also achieve high levels of combustion efficiency.

Discussions with Western companies and Russian power engineering factors have shown a perceived technical capability in the former USSR to assimilate many of these technologies, although such transfer is probably best achieved in a modular and phased approach, with increased investment in workplace-based quality management systems. The major factors influencing the transfer of these technologies, however, are all related to the macro-economic and political environment of the former USSR, and the influence of these on commercial practices in that region. Until these practices become more closely aligned to those encountered in the West, the pace of transfer of these technologies is likely to be slow in most industrial sectors including power engineering.

Exceptions to this general rule may be those power stations located near to Western borders where markets exist for electricity sales or environmental concerns are high; or in those regions where improved efficiency of generation may provide fuel savings and increased potential for Western exports and materials received as payment in kind. For coal-fired power stations east of the Urals, however, investment in environmental improvement is likely to require higher levels of Western multilateral support.

At present, however, almost all indigenous research and development appears to have come to a halt in the power engineering sector as a consequence of reductions in central funding for research, design and development; and a lack of sufficient financial resources in the privatised companies as a consequence of the necessity of ensuring financial survival.

The role of multilateral support therefore becomes even more critical, if indigenous power engineering research and development skills are not to disappear completely from the former USSR.

FUTURE RESEARCH PRIORITIES

Power Generation and Technology Transfer

The research described in this book has provided information on plans for new capacity in the power generation sector, and Western technologies being transferred for increased generation efficiency and reduced atmospheric emissions. At present, however, it appears that the construction of new power station capacity is progressing slowly in the former USSR as a consequence of political and economic uncertainties, indebtedness, and lack of investment funds. Research should consequently be continued to monitor progress in the assimilation of new and refurbished power station capacity in the former Soviet region, paying attention to possible changes in ownership which may occur as a consequence of power station indebtedness to fuel providers. The role of Western companies in these investment activities should also be monitored, paying particular attention to the technologies being implemented and the mechanisms used to achieve these transfers.

Atmospheric Emissions from other Industrial Sectors

Further studies should be made of SOx and NOx emissions from sectors other than electricity generation such as coking, ferrous and non-ferrous metals, transport and domestic heating. Coking ovens and the use of coke in the smelting of ferrous metals, emit high levels of sulphur dioxide; and the ferrous and non-ferrous metals industries also emit high levels of this pollutant (Peterson, 1993, pp. 30-42) during the production of carbon based electrodes and metal extraction from high sulphide ores. The transport industry, although not a large emitter of SOx pollutants, accounts for a particularly high proportion of NOx emissions, and domestic heating also accounts for high levels of emission of both SOx and NOx pollutants.

The methodologies of literature survey and case study described in this book could also be extended to a study of these other sectors.

'Clean Coal' Combustion Technologies

In view of the anticipated important role to be played by coal in the long term, particularly in Siberia, the Far East and Ukraine because of their comparatively high levels of coal reserves and a shortage of gas in Ukraine, it is apparent that development and pilot production should be commenced for the implementation of 'clean' combustion technologies using solid fuels in those regions. These technologies include fluidised bed combustion which enables SOx emissions to be reduced through the introduction of calcium compounds, and lower NOx emissions to be achieved as a consequence of lower combustion temperatures; and integrated gasification combined cycle systems (IGCC), in which coal is gasified to provide fuel for a gas turbine, and also as auxiliary heating in a heat recovery steam generator.

There is still significant development work to be done on these systems, however, in view of current operational problems associated with vibration and corrosion. It is important to note the long term importance of such development work in clean coal technologies, even though these developments may take place in Siberian locations which are a long distance from Europe.

Options for Risk Reduction

The research carried out in this project has demonstrated a general reluctance by Western financial institutions to advance loans for projects related to power generation in the former USSR, because of perceived high levels of political and economic risk in the region, and a shortage of domestic funds for significant initial investments. This research has also demonstrated, however, how business can be developed through the use of appropriate trading opportunities, and further research should consequently be commenced on options for the reduction of trading and investment risks in the former Soviet region.

CO_2 Emissions

The majority of the research project described in this book has been concerned with the transfer of technology for the reduction of SOx and NOx emissions. In view of the environmental influence of CO_2 emissions, particularly on the 'greenhouse effect', it is apparent that similar research should also be conducted on available options for the reduction of these emissions from the former USSR. Furthermore, the methodologies of literature review and case study used in this present research could also be extended to investigations related to CO_2 emissions.

Since CO_2, SOx and NOx emissions are all influenced by the selection of fuel and combustion processes, many of the conclusions from this present research will also have validity to the reduction of 'greenhouse gas' emissions, although further research may be required on the detrimental effects of methane emissions.

Tradeable Emissions

Several major Western countries are currently viewing the introduction of tradeable emission systems as a means to reduce SOx and CO_2 emissions, since governmental regulation can thereby be reinforced with market mechanisms which may also be more effective than fuel taxation based on pollutant content. This Western interest follows the apparent success of the 'tradeable permit' system in reducing SOx emissions from some 445 US power generation utilities from 10.5M tonnes in 1980 to 5.3M tonnes in 1995, although other advantageous cost factors may also have influenced the economic benefits of the programme (McLean, 1996).

A tradeable emission permit system would probably be extremely difficult to introduce in the former USSR, however, in view of the difficulties in assessment of the levels of atmospheric emissions described in Chapter 2, and the economic problems of transition described in Chapter 7. Nevertheless, it is important to investigate the feasibility of such a system in current political and economic conditions in the CIS, in case of future international political pressure to adopt such a mechanism.

Opportunities may also exist, however, for the trading of CO_2 emission permits on an international level to enable 'joint implementation initiatives' between OECD countries and the former USSR. As emissions from the former Soviet region have been reducing due to fuel switching and industrial decline (see Chapter 7), there may be scope for the sale of CO_2 emission permits to OECD countries, and the income thereby obtained can then be used for investment in newer and more efficient power stations. Such investment should also give further impetus for the reduction of atmospheric emissions.

As outlined in the previous paragraph but one, however, it is particularly important to investigate the feasibility of joint implementation initiatives in the transitional CIS economies, especially as internationally driven trading pressures could accelerate the pace of power station closures. Although such closures may not necessarily affect local electricity provision as supply is available through the national grid, local heating supplies from power stations could be affected, thereby causing problems for up to one third of the population.

NOTES

1. See US Bureau of the Census, *Statistical Abstract of the United States: 1994 (114th edition)*,Washington, DC, 1994 for US data on US power generation data and fuel mix (51 per cent of American electricity generated form coal in 1980). Data on US SOx and NOx emissions, and Soviet production of electric power are provided in Chapter 2 above.

REFERENCES

Gorlov, A.M. (1995), 'The Helical Turbine: A New Idea for Low-Head Hydro', *Hydro-Review*, Vol. 14, No.5 (September), pp. 1-4.

Hill, M.R. and McKay, R. (1988), *Soviet Product Quality*, London; Macmillan.

Hill, M.R. (1991), *Soviet Advanced Manufacturing Technology and Western Export Controls*, Aldershot: Avebury.

McLean, B. (1996), 'Political Evolution of the US Tradeable Permit System', Paper presented at the Conference on Controlling Carbon and Sulphur: International Investment and Trading Initiatives, Chatham House, London, 5-6 December 1996.

Peterson, D.J. (1993), *Troubled Lands: The Legacy of Soviet Environmental Destruction*, Boulder, CO: Westview Press.

Index